# 50 Amazing Things To See With Your New Telescope

By Allan Hall

Copyright © 2018 by Allan Hall

10 9 8 7 6 5 4 3 2 1

All rights reserved. No part of this publication may be reproduced, distributed, or transmitted in any form or by any means, including photocopying, recording, or other electronic or mechanical methods, without the prior written permission of the publisher, except in the case of brief quotations embodied in critical reviews and certain other noncommercial uses permitted by copyright law. For permission requests, write to the publisher, addressed "Attention: Permissions Coordinator," at the address below.

Allan Hall
1614 Woodland Lane
Huntsville, TX 77340
www.allans-stuff.com/50amazing/

Although the author and publisher have made every effort to ensure that the information in this book was correct at press time, the author and publisher do not assume and hereby disclaim any liability to any party for any loss, damage, or disruption caused by errors or omissions, whether such errors or omissions result from negligence, accident, or any other cause.

Any trademarks, service marks, product names or named features are assumed to be the property of their respective owners, and are used only for reference. There is no implied endorsement if we use one of these terms.

All images/graphics/illustrations in this book are copyrighted works by Allan Hall, ALL RIGHTS RESERVED, used with permission, or in the public domain.

Acknowledgements:

As always I have to thank my wife, Sue Ann, because without her prodding and help, my books would either never get done or would be far less than they are.

I would also like to thank my readers. Their emails, letters and reviews let me know that my work is appreciated which drives my desire to create more and better books.

# Table of Contents

1: Introduction ................................................................. 1

2: How to use this book ................................................. 3

3: The Objects ................................................................. 9

   3.1: Solar System ....................................................... 11

      #1: The Moon ..................................................... 13

      #2: The Sun ........................................................ 17

      #3: Saturn ........................................................... 20

      #4: Jupiter .......................................................... 22

      #5: Asteroids (Ceres and Vesta) ........................ 24

      #6: Iridium Flares .............................................. 25

      #7: International Space Station ....................... 27

      #8: Meteor Showers .......................................... 28

      #9: Comets ......................................................... 31

   3.2: Nebula ................................................................ 33

      #10: The Orion Nebula M42 ............................. 35

      #11: The Lagoon Nebula M8 ............................ 37

      #12: The Eagle Nebula M16 ............................. 39

      #13: The Omega Nebula M17 .......................... 41

      #14: The Dumbbell Nebula M27 ..................... 43

      #15: The Trifid Nebula M20 ............................. 45

      #16: The Ring Nebula M57 ............................... 47

      #17: The Crab Nebula M1 ................................ 49

      #18: The Ghost Of Jupiter NGC3242 ............... 51

## 3.3: Galaxies .................................................................. 53

#19: The Andromeda Galaxy M31............................... 55

#20: The Triangulum Galaxy M33 ............................. 57

#21: The Whirlpool Galaxy M51................................ 59

#22: The Sunflower Galaxy M63 ............................... 61

#23: Bode and Cigar Galaxies M81 & M82 .............. 63

#24: The Sculptor Galaxy NGC253 ........................... 65

#25: Centaurus A NGC5128........................................ 67

#26: The Virgo Supercluster ...................................... 69

#27: Our Milky Way..................................................... 72

## 3.4: Star Clusters ......................................................... 73

#28: The Pleiades M45 ................................................ 75

#29: The Great Globular Cluster In Hercules M13 ............. 77

#30: The Wild Duck Cluster M11............................... 79

#31: The Beehive Cluster M44 .................................. 81

#32: The Scorpius Globular Cluster M4................... 83

#33: The Pegasus Globular Cluster M15 ................. 85

#34: The Double Cluster in Perseus NGC869 & NGC884 ... 87

#35: Omega Centauri NGC5139 ................................ 89

#36: The Jewel Box NGC4755..................................... 91

#37: 47 Tucanae NGC104............................................ 93

#38: The Hyades Cluster C41 ..................................... 95

## 3.5: Stars......................................................................... 97

#39: Supernovae .......................................................... 99

  #40: The Double Double (Epsilon Lyrae) .......................... 101

  #41: The Double Stars of Albireo ..................................... 103

  #42: The Double Stars of Almach .................................... 105

  #43: The Double Stars of Castor ...................................... 106

  #44: Constellations ........................................................ 108

  #45: Polaris (The North Star) ......................................... 109

  #46: The Southern Cross ................................................ 112

  #47: Stellar Spectroscopy ............................................... 113

 3.6: Events ........................................................................ 115

  #48: Conjunctions, Occultations, Transits ....................... 117

  #49: Solar Eclipses ......................................................... 119

  #50: Lunar Eclipses ........................................................ 122

**4: The Constellations** ......................................................... 125

 4.1: Constellation Charts .................................................. 126

**5: Where to go from here** .................................................. 131

 5.1: Best months to view objects in the book .................... 132

 5.2: Index ........................................................................ 133

 5.3: Glossary ................................................................... 136

 5.4: Other books by the author ........................................ 160

 5.5: Notes ....................................................................... 166

This page intentionally left blank

# 1: Introduction

So you have this brand new telescope and you want to see some amazing things, but what exactly can you see? Sure, we all know you can see the moon and a bunch of stars. There has to be more than that, right?

Absolutely!

With even the most basic of telescopes you can see things like other planets, nebulae, galaxies and much more. The more advanced the telescope, the more objects you can see.

There are lots of great lists of objects such as the Messier, Caldwell, Herschel and more. Can you actually see and enjoy all those objects with just a basic inexpensive telescope? Not really.

I took the best items from those lists and added the best solar system items and more to make a list of objects that you can not only see with your new telescope, but actually enjoy.

The list is made up of astronomical objects that most people actually want to see. After all, you bought the telescope to look up at the sky and see things you can't really see without it, right? I know I did.

I decided to have the print book published in black and white simply because virtually everything you see in the night sky will be black, white, gray or maybe a little greenish. The bright colors you see in pictures and some books simply cannot be seen with amateur visual equipment but instead require long exposure astrophotography to capture.

## 50 Amazing Things to See With Your New Telescope

Since printing the book in color would easily triple the cost of the book and offer no real benefit, I chose to save you the money and make the book more affordable.

What objects you can see, and how well, depends on a lot of things including how much light pollution is in your sky, how experienced you are, and how good of a telescope you purchased. Do not get discouraged if there are some objects you simply cannot see with your telescope.

I have been doing this a long time, have spent enough on telescopes to buy a nice car, and there are still things I can't see at all. I can however see all the objects in this book.

## 2: How to use this book

Most objects in the book have three parts: an image of the object, a diagram showing the best months to view the object, and finally a star chart showing where the object is.

The image of the object in most cases is **not meant to be representative of what you will see in your telescope**. Exceptions to that include the sun and moon, and possibly Saturn and Jupiter if you have a sufficiently good telescope.

In the image set above, the image on the left is an image like what would appear in this book, the image on the right is what you might see in an amateur telescope.

The reason I include images that you will not be able to see is so that you know what you are looking for. I find that if I know what an object is supposed to look like, I can see a little more than if I just have no idea at all. In this way, unrealistic images can be beneficial.

Use the image set above and see if you can see more in the image on the right by seeing what is supposed to be there.

The diagram showing the best month is set for the central United States at midnight on the 15<sup>th</sup> of the central month. If you are in a different time zone then you can adjust these charts to suit you. Once you know how many months to add or subtract to one chart, you know for all of them.

Jan    Feb    Mar    Apr    May    Jun    Jul    Aug    Sep    Oct    Nov    Dec

Looking at the above chart, the object is highest in the sky on the 15<sup>th</sup> of April (as opposed to the 15<sup>th</sup> of March or 15<sup>th</sup> of May) at midnight CST and so April is the central point of the gray bar at the bottom.

Depending on your location, the object may not be far enough above the horizon to view on months other than the central month.

Unfortunately there is no perfect way to make such a chart but this one has proven to be fairly useful to a great number of people for many years so I continue to use it.

Lastly is the star chart. I have experimented with a lot of different types of charts in my books over the years and decided to try something new. These charts are screenshots out of the freeware program Stellarium and were used with the permission of Alexander Wolf.

I have framed the object with some constellations and major stars and then put a huge arrow pointing to the object you are looking for.

After many years of trying to show people where objects in the sky were it became clear the easiest way to accomplish this was to have them find the constellations, and then navigate from there.

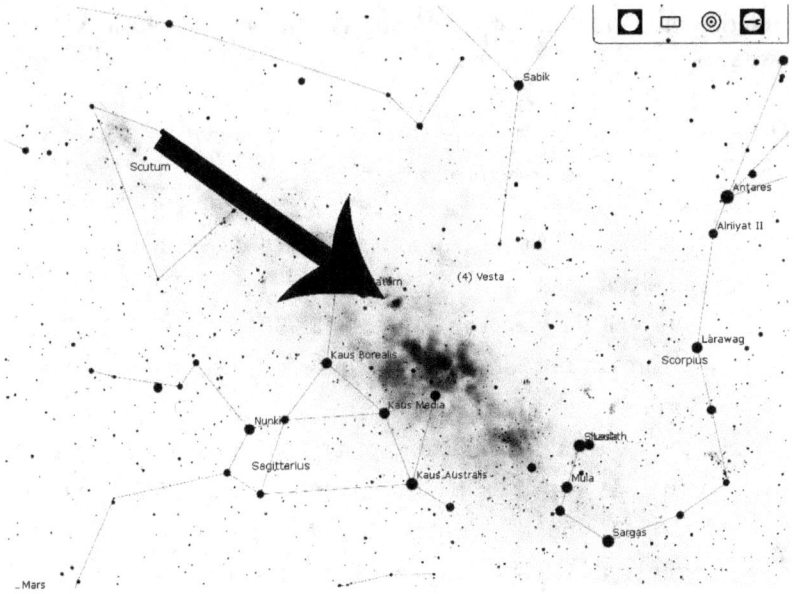

Because books are printed on imperfect presses, and the paper and ink are never perfect, the images can be less than ideal. For that reason you can download the charts off of my website at www.allans-stuff.com/50amazing/ in full resolution and color. Of course you could also download Stellarium from https://stellarium.org/ and look up the objects yourself.

I suggest you buy a good planisphere such as the Miller Planisphere so you can easily find the constellations. This will make finding all the objects in this book fast and easy, even for the novice.

Planispheres typically take a novice a few minutes to figure out and then they are good to go. A quality one like the Miller's will work excellent outside in the dew and last years. They are a great investment. You can find them at https://amzn.to/2NqrfVn

If you prefer not to purchase a planisphere, look at the website https://in-the-sky.org/whatsup.php as it will provide much of the functionality and more, it is just not as convenient to have outside under the stars.

Once you get outside and start to look for objects there are some important things you need to remember:

1) Let your telescope warm up or cool down. If it has been inside in Texas in the summer it is probably 70 degrees and you will be taking it outside into 90 degree temps. Let it acclimate outside for an hour before use and your views will be much better.
2) Your eyes need time to adjust too! Find some place away from as much light as possible, spend 15-30 minutes goofing around (not looking at your phone or tablet) until your eyes are adjusted and then start viewing. You can see much more detail with adjusted eyes than without.
3) If you need a light to look something up or adjust something, use a red light. Also make sure the red light is as dim as possible. You want just barely enough light to do the task you need done so your eyes stay used to the darkness.
4) Be patient, it takes time and experience to make out the most of an object. While I don't expect everyone who reads this book to do this hobby for thirty years or more, the more experience you have, the more you will see, period. This also doesn't mean you won't see anything unless you have thirty years of experience, it just means you need to slow down and work at it.
5) Don't be afraid to ask for help. Find your local astronomy club (call a local college, high school, or library) and tell them you could use some help. Most club members are happy to help a newcomer to the hobby.

6) Can't find local help and having a problem? Check out my website at allans-stuff.com and my YouTube site at www.youtube.com/c/AllanHall.

On many of the objects you may see things on the end of the names such as M20, NGC104, or C41. These are the catalog numbers of the objects.

Virtually every object in the sky has a catalog number so that astronomers can reference specific objects. There are different catalogs, some of which are object type specific, for example some are only stars.

In this book I have tried to limit things to three popular catalogs; Messier (M001), New General Catalog (NGC001), and Caldwell (C001).

As you get more experienced and want to expand beyond the objects presented in this book you can find information on the other objects such as the Messier if you are in the Northern Hemisphere or Caldwell if you are in the Southern.

There are many more objects in the catalog. The Messier contains 110 objects while the Caldwell contains 109. The New General Catalog contains thousands.

There are of course many other catalogs you can choose from but that is far beyond the scope of this book.

One last note; you will notice that the glossary has terms included that do not appear in this book. This is intentional so that it can be used as a reference as your interest in astronomy grows beyond the confines of this beginner text.

50 Amazing Things to See With Your New Telescope

# 3: The Objects

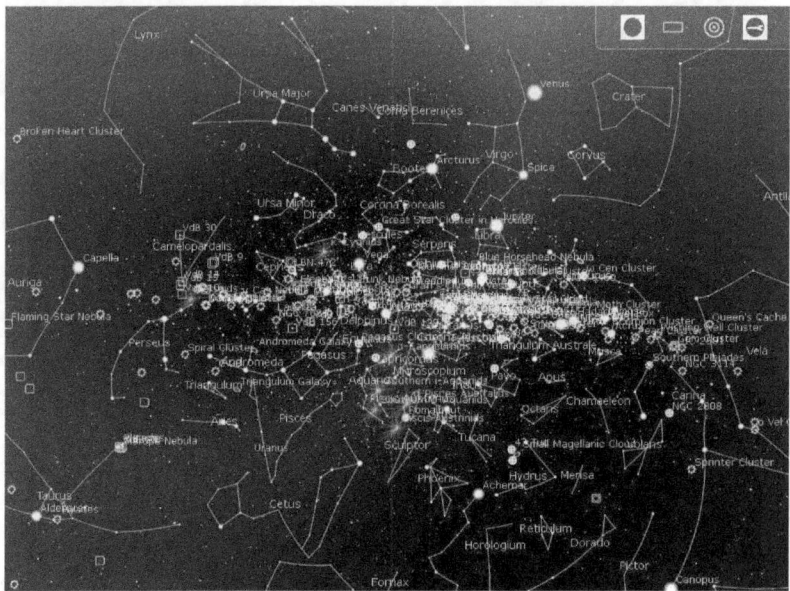

There are thousands and thousands of objects in the night sky. There are hundreds you can see that will show as more than just a dot on a black background. The following objects are the best of those objects that are viewable using small beginner telescope equipment.

50 Amazing Things to See With Your New Telescope

## 3.1: Solar System

Our solar system houses some of the best objects you can start viewing, and the only ones you can view in the middle of the day (sun and moon).

In fact, even if you have never used a telescope or binoculars you have already observed at least those two celestial objects so you have a head start you didn't even know about!

What better place to start your journey than with the objects that are closest to you.

50 Amazing Things to See With Your New Telescope

## #1: The Moon

OK, this one you can clearly see without a telescope, but when you use a telescope the level of detail you can see is simply….amazing. All of a sudden craters, valleys and mountains jump out at you. For the first time you might realize that this is not just a big glowing sphere in the sky but actually is an amazing object that shows a record of its creation right on the surface for you to see.

The moon is so interesting to most, that I even bought a telescope, camera and other specialized equipment just for use on it. Using these tools and specialized techniques it is possible to get even more detail out of the moon.

One of the most important things to remember when viewing the moon is that you do not want to view it when it is full. I know this seems counter-intuitive but the full moon will look flat and featureless.

You want to view the moon when it is partially lit. The line you see between the lit and unlit portions of the moon is called the terminator, and viewing close to this line is where you can see the greatest amount of detail. The reason for this is simple, shadows.

When the moon is full, sunlight hits the moon's surface straight on meaning that there are no shadows, and therefore no definition of the surface features.

Think about when you take a picture of someone with no light except that of your camera or phone right in front of the person. Now take that source of light and move it off to the side a little so that it is not shining right into the person's face and you get a much better picture. Again, this is due to adding shadows.

Another thing to watch out for with the moon is that it is very bright. You may have received a "moon filter" with your telescope for this very reason. These filters are just like sunglasses for your telescope blocking a good portion of the light from hitting your eye.

So why is the moon so bright? The moon's surface is mostly a light gray rock color (think concrete) which is directly lit by the sun. Just like if you were in a large concrete parking lot here on earth that reflected light can be blinding. This is made worse by your eyes being adapted to the dark since you are probably viewing the moon at night so the difference between what your eyes are adapted for and the brightness of the moon is very substantial.

Of course magnifying the surface also magnifies the light into your eye making that even worse!

While I have never heard of anyone having their eyes damaged by looking at the moon through a beginner telescope without a filter, I have had many people express discomfort and not being able to see for a while after looking away from the telescope.

If you do not have a moon filter and you experience any discomfort I would suggest you get one, they are typically very economical and screw right on to your eyepiece (make sure your eyepieces have threads before you order a filter).

Once you spend some time looking at the moon you will be able to see a lot of detail. Just like anything in astronomy, the more you do an activity, the better you get.

While the moon is the brightest nighttime object you are likely to view, you can still see more detail as you gain more experience.

## #2: The Sun

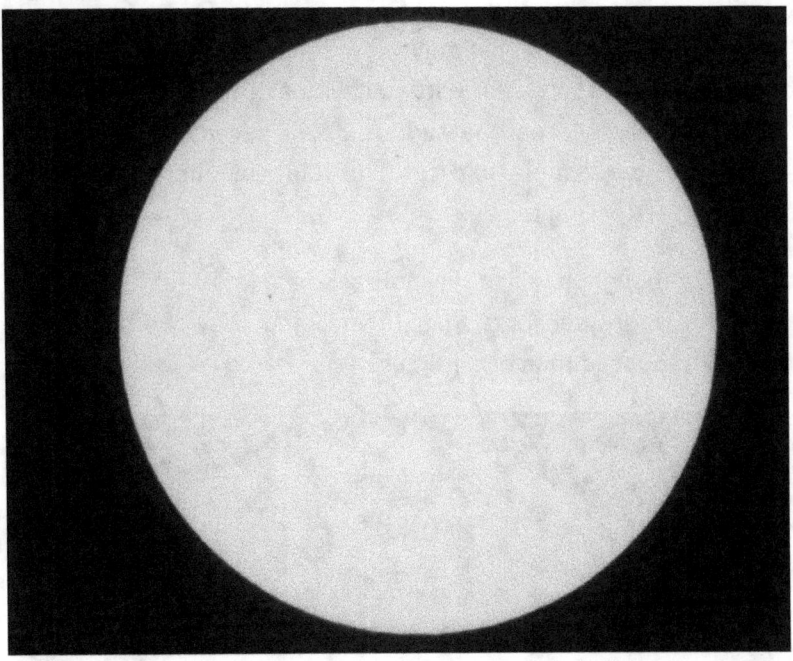

**The sun should only be viewed by using a professionally made solar filter that goes over the end of your telescope. Never look at the sun without this professional protection. Be sure to cover any finders as well.**

Many people have tried things like welder's glass, multiple sheets of window tint, and other creative ways to make themselves permanently blind. Personally, I like my vision, and it is just not worth risking my eyesight to save a few dollars.

Solar filters come in two basic types; glass and film. Film are cheaper, glass tend to provide better images and last longer. I have used both and tend to use some film models when I want something to use once or twice, and I purchase glass models for use with telescopes if I plan on using them for more than that.

A simple solar filter will typically allow you to see quite a bit on the sun such as sunspots and a little activity at the edge of the solar disk. You will not see any serious detail because that requires specialized telescopes that cost substantially more than most beginners are willing to pay.

That's OK though, tracking sunspots can be a lot of fun. Sometimes there are more sunspots than others, and they are sometimes more pronounced than other times.

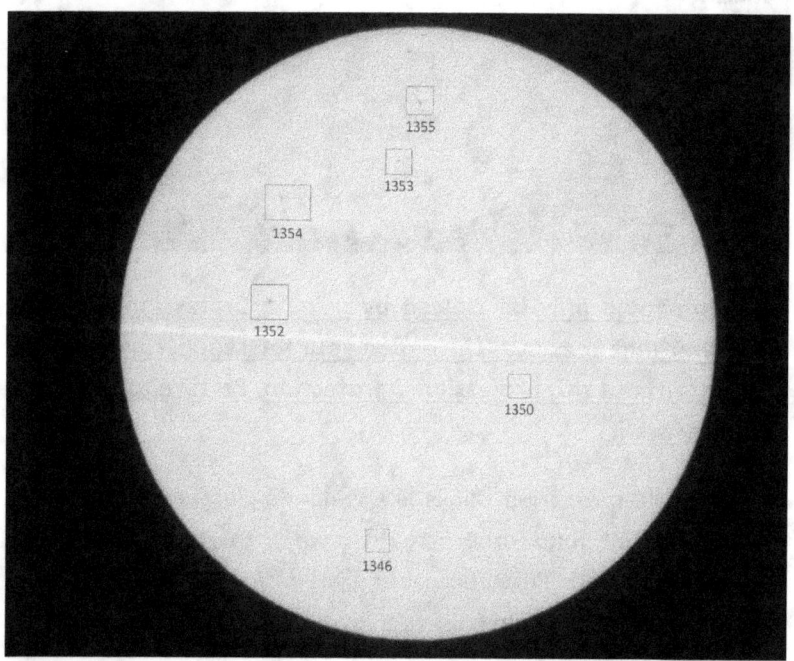

You might go to www.spaceweatherlive.com/en/solar-activity/sunspot-regions to see what sunspots should be visible each day and compare that to what you see in your telescope.

## 50 Amazing Things to See With Your New Telescope

One thing that may seem a little strange is that viewing the sun can actually be a little challenging. Once you put your solar filter on your telescope everything you view in the telescope will be pitch black, until you get the sun in your field of view. This is a lot harder than you might think, particularly since you cannot use your finder scope to help you.

There are a couple ways to assist you in getting the sun in your telescope.

The easiest way is to get the eyepiece with the least amount of magnification you have, this will be the one with the highest number. If your telescope came with a 25mm and a 10mm, use the 25mm.

Now stand in front of the telescope and a little to the side so that you can see the shadow of the telescope on the ground or the side of a building.

Move the telescope around until you get the smallest shadow possible and then look through the eyepiece. You may need to move the telescope around a little but you should be close.

Also make sure that you adjust the focus because if your telescope is far enough out of focus, you will not be able to see the sun even if you are pointed right at it. Usually moving the focus all the way one direction and then all the way the other direction quickly will tell you in a few seconds if you are on target or not.

The best method of getting the sun in your telescope is to purchase a solar finder like the one from Televue. These run about $30 but will pay for themselves the first time you find the sun in 30 seconds while the person next to you fights for half an hour.

# #3: Saturn

Saturn is usually the first planet people see in a telescope, and the first object that really makes people say "WOW!" I know I said that and a whole lot more while I was jumping up and down with glee (and no, I was not a child, I was a full grown adult living on my own when that happened).

There are a lot of things about Saturn that really get your attention, the first of which of course is the rings. Once someone sees the rings, there is no doubt in their mind that they are looking at another planet.

With Uranus, Neptune, Mars, Venus, and Mercury in beginner astronomy equipment you see a bright dot. Yes, Mars is a reddish bright dot, but still just a dot. Who can say you are looking at a planet and not something else, maybe a star?

When you see Saturn, there is no doubt. In even the most basic of equipment you can see the ring around a round object. It is clearly too large to be a star as all the stars are pretty much just dots, and this is much larger than that.

With slightly better equipment you can see the division in the rings called the Cassini Division and you might also start to see some striations in the clouds on the planet.

Moving up to a little better equipment you will be able to see multiple shades of the different rings, clear variations in the cloud cover of the planet, and some of the planet's moons.

This is an object you can come back to time and time again and see something more incredible than the last time you looked. You can even get specialized equipment for viewing and imaging planets to get even more detail out of this amazing object.

To find Saturn, go online to https://in-the-sky.org/data/planets.php or use your favorite software such as Stellarium.

## #4: Jupiter

Jupiter is not quite as awe inspiring at first as Saturn, but it certainly can make you spend hours staring at it just the same.

What it lacks in a ring like Saturn's, it makes up for with amazing patterns in the cloud cover, the big red spot (which you may or may not be able to see depending on when you view it, and the equipment you have), and the easily found moons.

Just like Saturn, the better equipment and more experience you have the more you can see. You can also buy specialized planetary equipment and get even more detail out of it.

Even without extra specialized stuff, Jupiter is a fantastic target. I have a lot of fun matching the moons up with what I see in my astronomy program. Using software or charts is the only way to identify which moons are which because as little dots they will appear to be in your telescope, there are no identifying colors or features.

The above image was the first image I took with my "real" astrophotography telescope and a DSLR camera. You can clearly see Jupiter and three moons. This is actually a fairly good representation of what you can expect to see with a reasonable telescope under dark skies visually.

To find Jupiter, go online to https://in-the-sky.org/data/planets.php or use your favorite software such as Stellarium.

# #5: Asteroids (Ceres and Vesta)

Ceres on the left, Vesta on the right, not to scale. Image credit: NASA / JPL-Caltech / UCLA / MPS / DLR / IDA / Justin Cowart

There are two asteroids that you can readily see with a small telescope, Ceres and Vesta. The image above is just for reference, what you will really see is more like a dot of reflected light.

These objects are located in the asteroid belt between Mars and Jupiter. Ceres is almost 600 miles in diameter and is mostly round, Vesta on the other hand is more American football shaped and approximately 320 miles long at its longest point.

While these are unlikely to be more than dots in your eyepiece, it is pretty neat to realize you are seeing asteroids. With slightly higher end equipment, more powerful eyepieces and darker skies you can see some shading on Ceres and the oblong shape of Vesta.

To find these asteroids and more check out https://in-the-sky.org/whatsup.php and select the asteroid tab.

# #6: Iridium Flares

Double Iridium flare from 6 and its replacement 51. Image by Jud McCranie

An iridium flare is not at all what it sounds like. There is no iridium element on fire as you might expect, instead it is simply a reflection off the solar panels of an iridium satellite.

Iridium satellites are a set of communications satellites orbiting the earth to provide telecommunications services in the 1-2GHz range. There are approximately 70 of these in orbit at any given time (66 active plus spares).

The original plan of the network of satellites called for 77 of them, which is the atomic number of iridium, hence the name iridium satellite.

As the sun sets, the angle of the sun and satellite's solar panels creates a very bright reflection back to earth. This is often seen as a very bright streak low in the sky lasting for a few seconds.

You can find out where the next iridium flare in your area will be by visiting www.heavens-above.com/IridiumFlares.aspx and looking at the chart.

Not only are these a lot of fun to see by yourself (there isn't that much to look at while the sun is still up), they are also a blast to get friends all looking at that area of the sky right before it happens. It is sure to illicit several oohs and ahhs.

## #7: The International Space Station

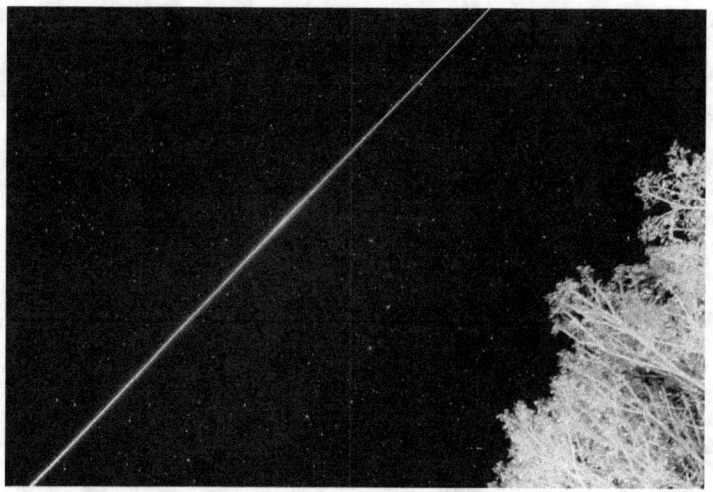

What you are likely to see when the ISS passes overhead

In all honesty, to see much more than a bright streak across the sky you will need to move your telescope really fast and/or use a video camera and capture a frame or two.

That being said, chasing the ISS can be a lot of fun! So many people think so that NASA created the Spot the Station website at spotthestation.nasa.gov where you can get dates and times for your location when you can see the ISS overhead.

# #8: Meteor Showers

Meteor showers are both something in our solar system, and an event (another section located later in this book) but I felt it belonged here more than there.

Meteoroids are chunks of rocks and debris flying through space. Meteors are meteoroids that are hitting our atmosphere and burning up. If the meteor reaches the earth, it is called a meteorite.

Comets can leave trails of meteoroids in space and as the earth travels through these areas of space, we "run into" these meteoroids and the ones we hit enter the atmosphere putting on a spectacular light show called a meteor shower.

The name of the meteor shower, such as the Lyrids, tells you where in the sky the meteors appear to come from (the constellation Lyra in this example).

As the meteors streak across the sky, if you draw a line connecting the point of origin in the sky of all the meteors you see, those lines should intersect somewhere in the named constellation.

There are eleven main meteor showers a year; Quadrantids, Lyrids, Eta Aquarids, Delta Aquarids, Perseids, Draconids, Orionids, Taurids, Leonids, Geminids, and Ursids. They happen on the same date each year so you can easily plan ahead.

The ones in the milder months of the year can be a lot of fun as you and your friends can lay out under the stars facing in different directions and count how many you see for later comparisons.

| Shower Name | When | Shower Name | When |
|---|---|---|---|
| Quadrantids | Early January | Lyrids | Late April |
| Pi Puppids | Late April | Era Aquariids | Early May |
| Arietids | Mid June | June Bootids | Late June |
| Southern Delta Aquariids | Late July | Alpha Capricornids | Late July |
| Perseids | Mid August | Kappa Cygnids | Mid August |
| Aurigida | Early September | Draconids | Early October |
| Orionids | Late October | Southern Taurids | Early November |
| Northern Taurids | Mid November | Andromedids | Mid November |
| Alpha Monocerotids | Mid November | Leonids | Mid November |
| Phoenicids | Early December | Geminids | Mid December |
| Ursids | Late December | | |

Chart of popular meteor showers.

## #9: Comets

Comets are basically balls of rock in our solar system which are being blown apart by the solar winds emanating from the sun. This effect is what causes the comet's tail you can see.

While very bright obvious comets are fairly rare, comets you can observe with binoculars or a small telescope are not. For a list of comets you can observe now take a look at the website in-the-sky.org/data/comets.php.

## 3.2: Nebulae

Nebulae are large areas of gas in the sky. This gas is usually a greenish gray color to the naked eye although some can exhibit other colors to certain people (people unusually color sensitive) under good seeing conditions.

Nebulae can be caused by a variety of things such as gas and debris ejected by the explosion of a dying star or gas being collected in areas where stars are forming.

There are two basic types of nebulae; reflection and emission.

Reflection nebulae are visible because they reflect the light of a nearby star.

Emission nebulae are visible because the gases themselves emit light. This emission is usually due to ionization caused by nearby stars.

## #10: The Orion Nebula M42

Jan  Feb  Mar  Apr  May  Jun  Jul  Aug  Sep  Oct  Nov  Dec

One of the best nebula you can observe is the Orion nebula located right smack in the middle of the constellation Orion. This nebula is so large and bright, you can see it by just looking up with nothing but your eyes in even semi dark conditions.

With binoculars or a small telescope the nebula really jumps to life with plenty of beautiful details and wisps of dust and gas.

While you can spend a lot of time viewing the details with even a pair of cheap binoculars, adding more equipment and experience can bring out details you would never believe you could see.

# 50 Amazing Things to See With Your New Telescope

This was the first nebula I saw with my first "real" telescope, a $350 114mm f9 Newtonian reflector on an equatorial mount. That same caliber of scope today would cost about $125 and provide about the same jaw dropping views I saw all those years ago.

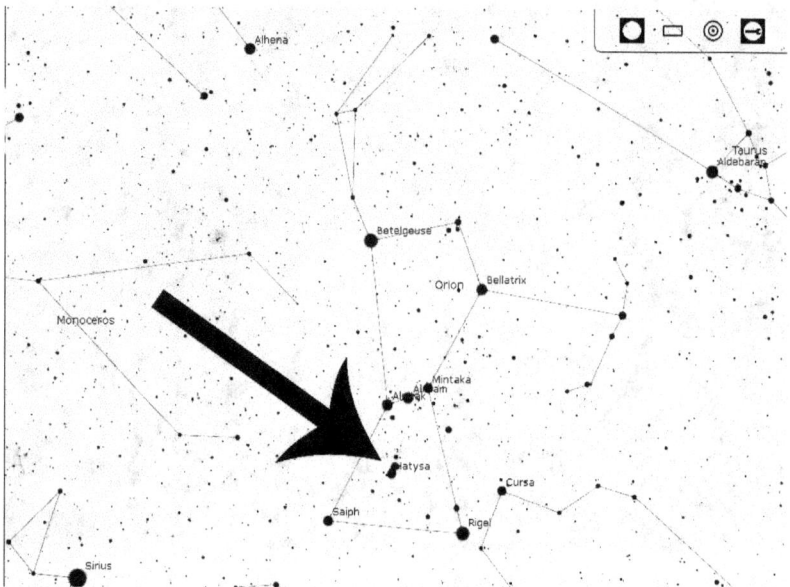

Finding the Orion nebula is very easy since from even moderately dark skies you can clearly see it with just your naked eyes.

If you can find Orion, find the three stars that make his belt.

Now find the three stars that form his sword hanging from the belt and you should see a fuzzy patch right in the middle, that is the Orion Nebula.

# #11: The Lagoon Nebula M8

Jan   Feb   Mar   Apr   May   Jun   Jul   Aug   Sep   Oct   Nov   Dec

The Lagoon nebula is a pretty bright nebula in Sagittarius that can be seen with the naked eye under dark conditions. With a pair of binoculars or a small telescope details start to pop out even under less than ideal conditions.

One thing that makes this object just outstanding is the beautiful open cluster right in the middle of it.

# 50 Amazing Things to See With Your New Telescope

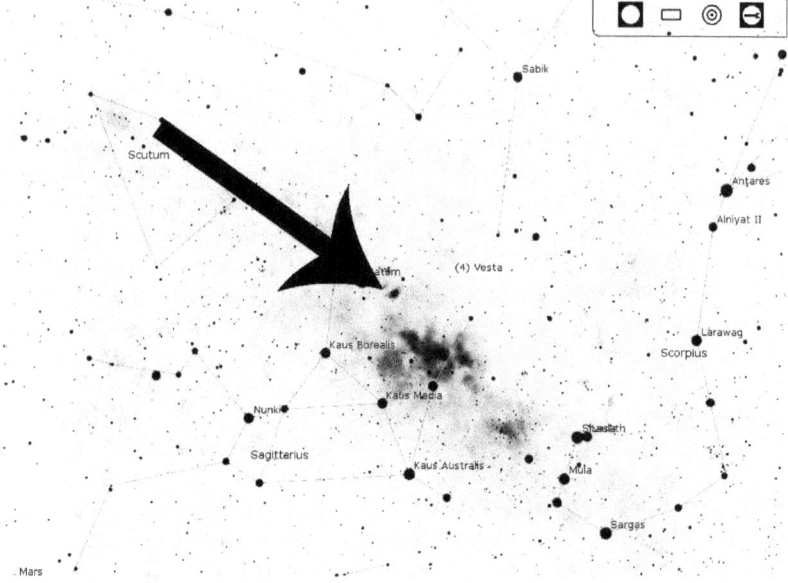

There is a lot of stuff in the area of the Lagoon nebula so finding something is not that difficult, finding the right something, that might prove a little more challenging.

M8 is just outside Sagittarius just south east of a line drawn from Polis to Vesta about half way between the two. This is almost smack in the center of the Milky Way.

## #12: The Eagle Nebula M16

Jan　Feb　Mar　Apr　May　Jun　Jul　Aug　Sep　Oct　Nov　Dec

If you have seen the famous pictures of the "pillars of creation", this is the stellar object where those pillars live. It gets its name from the fact that it looks somewhat like an eagle with its wings spread.

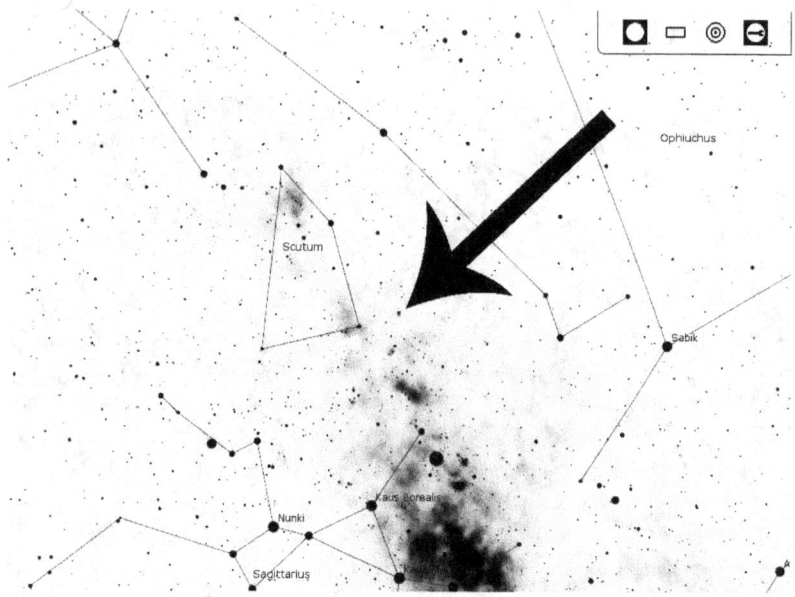

M16 is right in the same area as M8, but a little north of that position. It lies just west of the constellation Scutum, again right in the middle of the Milky Way.

## #13: The Omega Nebula M17

Jan  Feb  Mar  Apr  May  Jun  Jul  Aug  Sep  Oct  Nov  Dec

This is a very interesting object as even with the naked eye you can see the open cluster to the side. Add a pair of binoculars or a small telescope and you can see a nice patch of nebula to the side of that cluster.

If you add even nicer equipment, you will be treated to even more nebulosity as it seems to expand proportionally to the size of your telescope.

It really is a fun target.

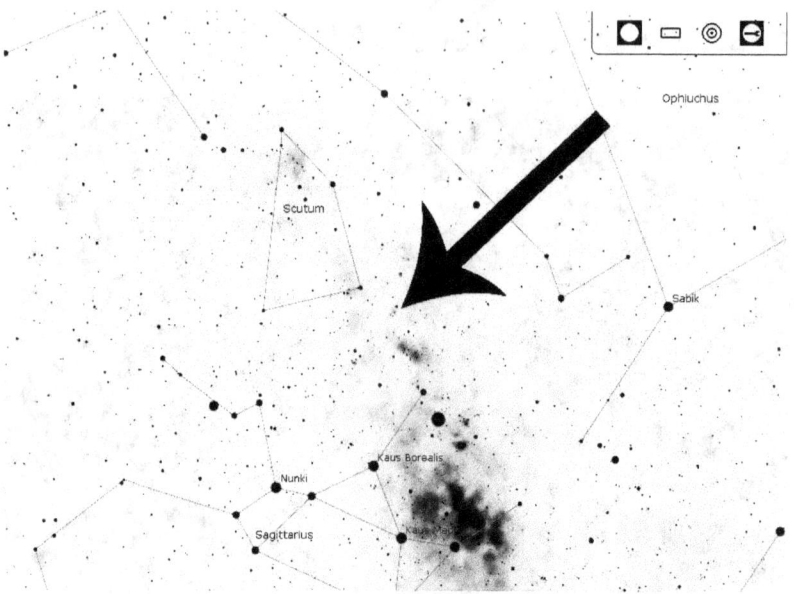

If you start at M16 just west of Scutum and head south towards Polis in Sagittarius, you will not have to go far before you find M17.

## #14: The Dumbbell Nebula M27

Jan   Feb   Mar   Apr   May   Jun   Jul   Aug   Sep   Oct   Nov   Dec

Visually, the Dumbbell nebula is a small open cluster with a little nebulosity around it. Small telescopes can make out the general bubble shape of the blue areas in the nebula (usually seen as green) under moderately dark skies.

The sides of the dumbbell which show up red in photographs are very difficult to see in smaller telescopes. If you have a moderate sized setup with very dark skies you might can catch a glimpse of them.

# 50 Amazing Things to See With Your New Telescope

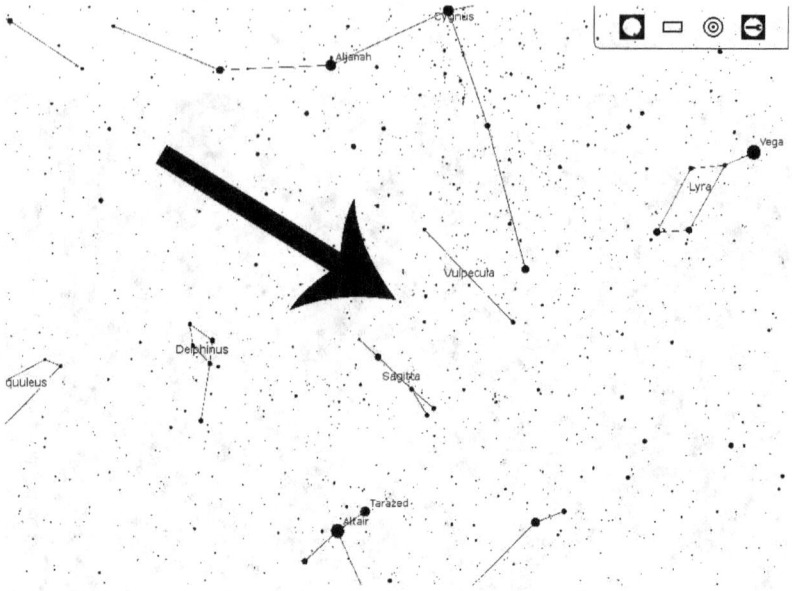

To find M27 just find the two parallel lines of Vulpecula and Sagitta, then look dead in between the two slightly to the north east.

## #15: The Trifid Nebula M20

Jan  Feb  Mar  Apr  May  Jun  Jul  Aug  Sep  Oct  Nov  Dec

This nebula is one of the most recognizable objects for most people as it is both a reflection and emission nebula in one. This results in both a red and blue nebula in one, although most people cannot see this visually.

Even without the colors, and really even without being able to see any of the nebula, the open cluster surrounding this area is still nice to look at.

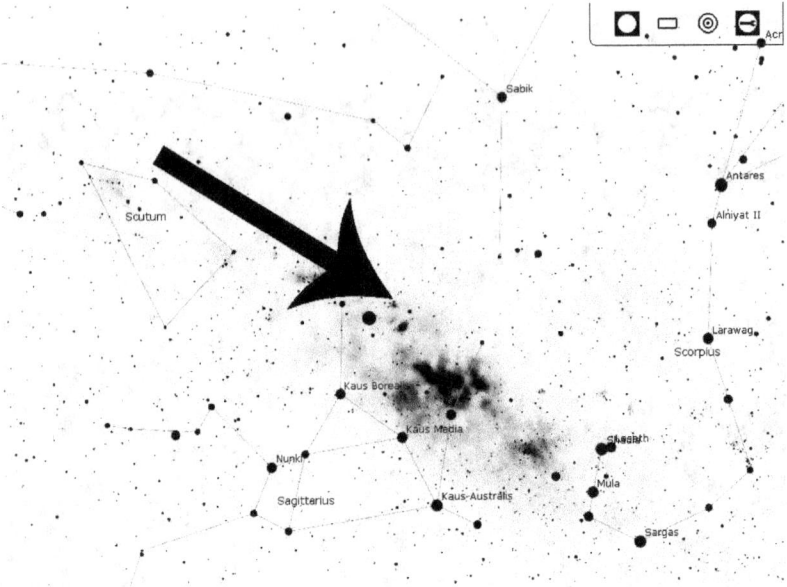

M20 is just outside Sagittarius right on a line drawn from Polis to Vesta about half way between the two. This is almost smack in the center of the Milky Way. This is just a tad south of M8.

# #16: The Ring Nebula M57

Jan  Feb  Mar  Apr  May  Jun  Jul  Aug  Sep  Oct  Nov  Dec

This was the first nebula I viewed with my first "really serious" telescope. It was also the first object where I saw colors other than the standard gray or green as this nebula was decidedly blue.

In photographs this object can seem red, white, and blue from the outside to the inside. Visually it typically presents as a green or blue doughnut.

# 50 Amazing Things to See With Your New Telescope

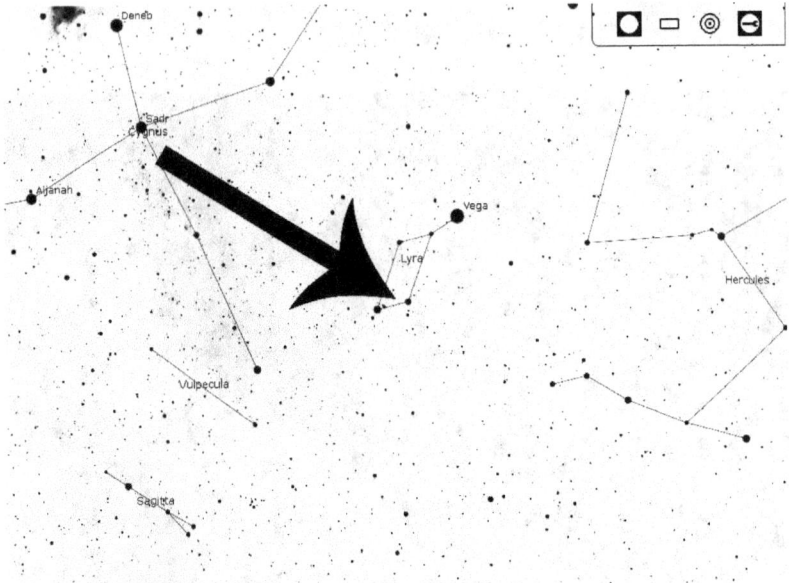

Vega in Lyra is one of the brightest stars in the sky so it should be easy enough to find and is the "top" of Lyra.

Now look at the other end of Lyra in between the two "bottom" stars and you will find M57 about two thirds of the way between Sulafat and Sheliak just a tiny bit south of the line between the two stars.

## #17: The Crab Nebula M1

Jan  Feb  Mar  Apr  May  Jun  Jul  Aug  Sep  Oct  Nov  Dec

With a small telescope you can just make out this supernovae remnant in the constellation Taurus. It tends to look like a small puff of smoke.

Larger telescopes can start to show a little structure but no real color of any kind.

# 50 Amazing Things to See With Your New Telescope

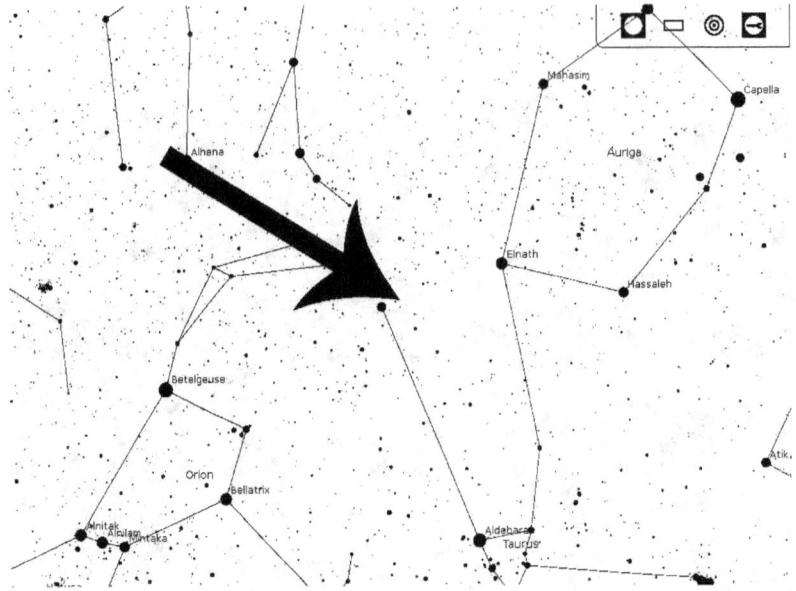

M1 is located just off the star Zet Tau, in the constellation Taurus of course, in the south western part of the constellation.

# #18: The Ghost of Jupiter NGC3242

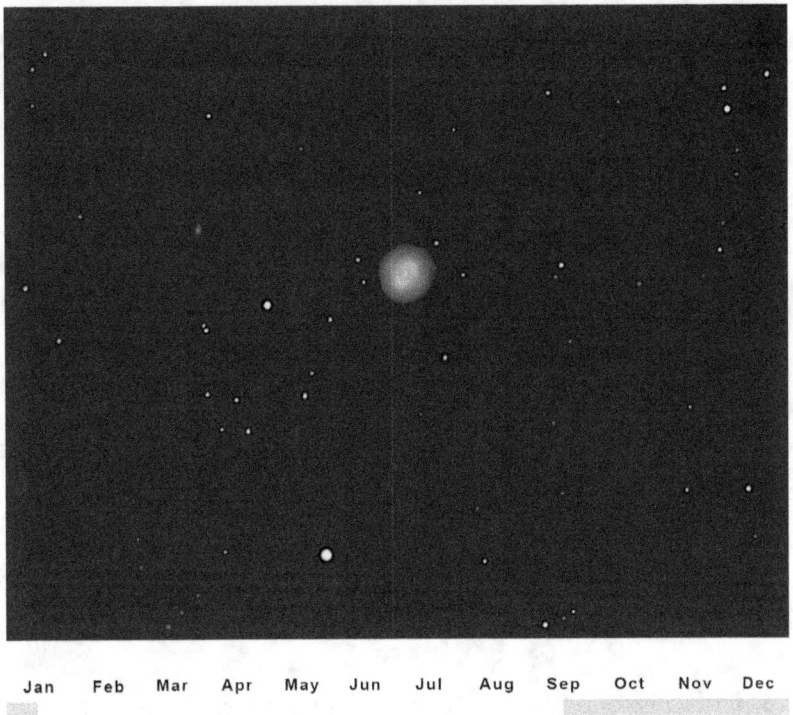

Jan  Feb  Mar  Apr  May  Jun  Jul  Aug  Sep  Oct  Nov  Dec

This planetary nebula is approximately the size of Jupiter in the sky so you are unlikely to get a lot of detail in a small telescope, but it is a lot of fun seeing how much you can coax out.

This object is basically a dot (white dwarf star) surrounded by a nebula, which in turn is surrounded by a fainter nebula.

Small telescopes can see the object and under good conditions and reasonable magnification can make out the central star and its halo.

Stepping up to a midsize or larger telescope in a moderately dark site will let you start to see the outer halo.

The image above was done with a large 10" SCT telescope and should give you an idea of what you are looking at but you will be able to see very little of this structure.

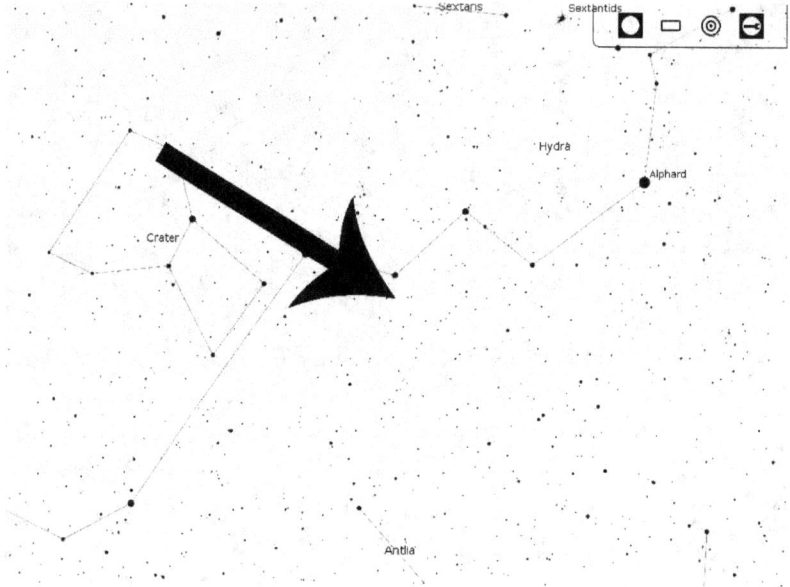

To find this one look virtually in the center of the massive constellation Hydra, right next to the star Mu Hydrae.

## 3.3: Galaxies

Yes, galaxies are really far away, but they are also massive which allows us to see some of them. Depending on how far away and how they are oriented, you might just see what looks like a fat star, or you might see the entire spiral galaxy in all its glory.

Many people spend a lot of time and effort trying to coax out a lot of detail from these far away cities in space.

50 Amazing Things to See With Your New Telescope

# #19: The Andromeda Galaxy M31

Jan   Feb   Mar   Apr   May   Jun   Jul   Aug   Sep   Oct   Nov   Dec

Our closest neighbor in the universe is the Andromeda galaxy, it is also the galaxy that will soon impact and devour our galaxy whole.

Now before you run out and cash in your life insurance policy, soon in this context means in galactic terms, or millions of years. Even when it does happen, it is unlikely that it will affect our solar system at all because there is so much empty space between solar systems in both the Andromeda and Milky Way galaxies.

# 50 Amazing Things to See With Your New Telescope

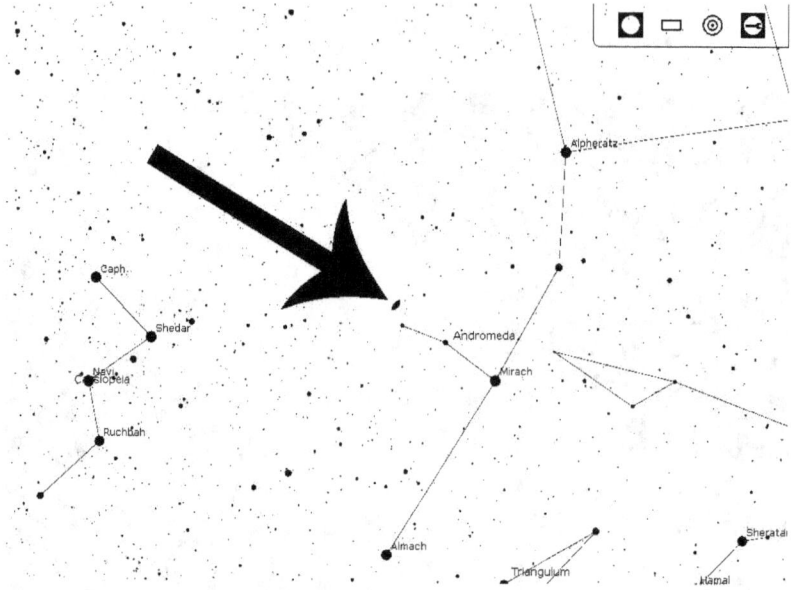

M31 is extremely bright and lies right next to the star Nu And in the constellation Andromeda. If you can find the constellation, the fuzzy bright patch should become obvious quickly.

## #20: The Triangulum Galaxy M33

Jan   Feb   Mar   Apr   May   Jun   Jul   Aug   Sep   Oct   Nov   Dec

While this face on spiral galaxy can be seen with the naked eye, it is much nicer when viewed through a telescope.

From a moderately dark site with a medium sized telescope you can clearly make out the spiral arms. Smaller telescopes can certainly see the galaxy and make out some details such as the general shape.

# 50 Amazing Things to See With Your New Telescope

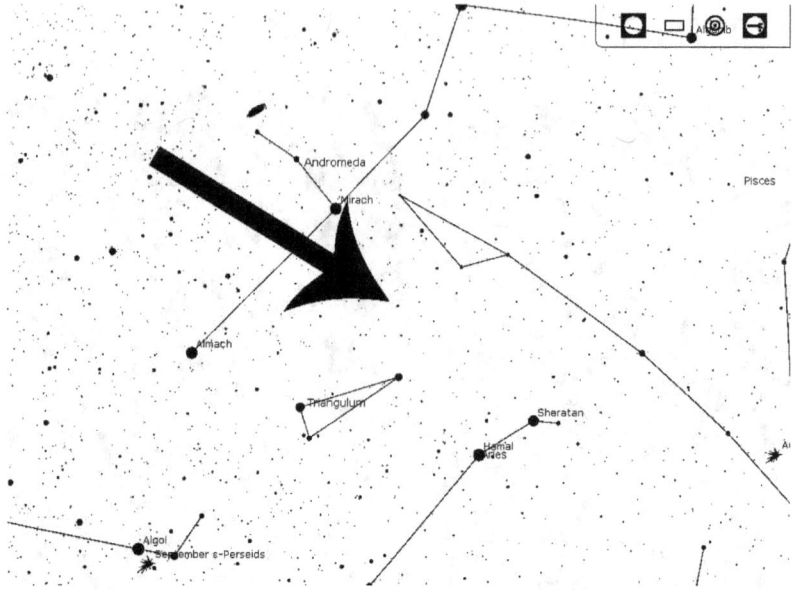

M33 lies almost in the center of Andromeda, Triangulum, and Pisces and should appear as a fuzzy patch in moderately dark skies with just the naked eye.

## #21: The Whirlpool Galaxy M51

Jan    Feb    Mar    Apr    May    Jun    Jul    Aug    Sep    Oct    Nov    Dec

This object is a real treat to view because it is another BOGO for you. Point your telescope into the Canes Venatici constellation towards this galaxy and you will quickly see not one, but two galaxies.

The larger member of this duo is a face on spiral galaxy that is in itself a beautiful galaxy. What is really interesting here is that it seems to be grabbing hold of a second galaxy, a dwarf galaxy, and pulling it "down the drain" so to speak. This is where the whirlpool name comes from.

Easily visible with a small telescope, or even binoculars, this one will keep you looking for years to come, regardless of how big your telescope gets.

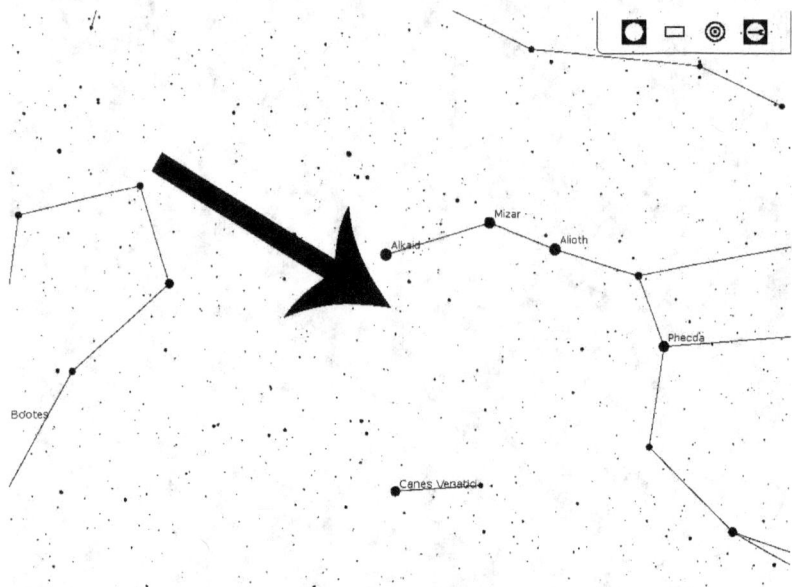

To find M51 simply find the tail of Ursa Major, the tip of which is the star Alkaid. Now draw a line to Canes Venatici's southernmost star, Cor Caroli and travel about a quarter of the way there from Alkaid.

# #22: The Sunflower Galaxy M63

Jan   Feb   Mar   Apr   May   Jun   Jul   Aug   Sep   Oct   Nov   Dec

This spiral galaxy is a little small for smaller telescopes. You can see it, but there is not much to it. It shows up as pretty much a really fuzzy star. For medium and larger telescopes however, this is an amazing galaxy so it is worth trying with whatever you have.

The really interesting thing about this object is that instead of having a few long arms like most spiral galaxies you see pictures of, it has a lot of very short arms that give it a very unique pattern that resembles a sunflower.

## 50 Amazing Things to See With Your New Telescope

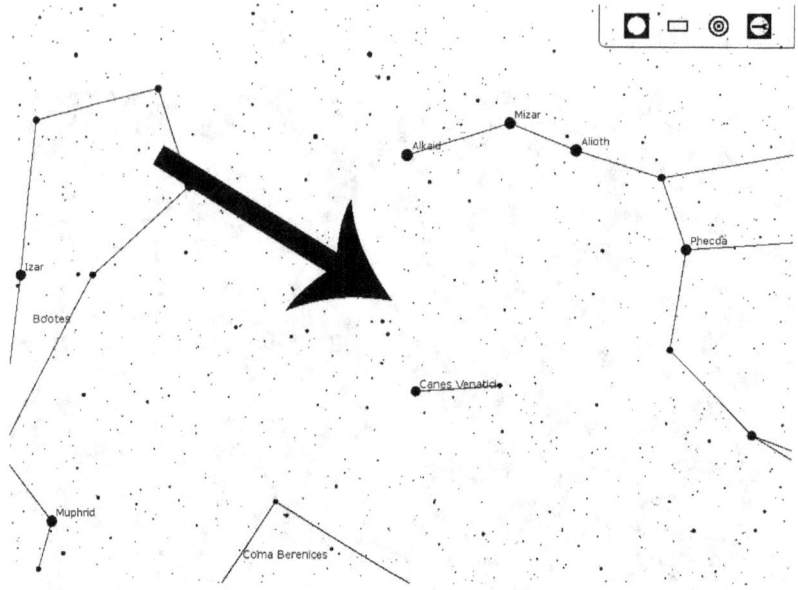

To find M63 simply find the tail of Ursa Major, the tip of which is the star Alkaid. Now draw a line to Canes Venatici's southernmost star, Cor Caroli and travel about three quarters of the way there from Alkaid.

## #23: Bode and Cigar Galaxies M81 & M82

This is another BOGO and you have both a spiral and starburst galaxy to look at. While visible in even binoculars, a small telescope will clearly show the shape of the galaxies while larger telescopes begin to show the structures.

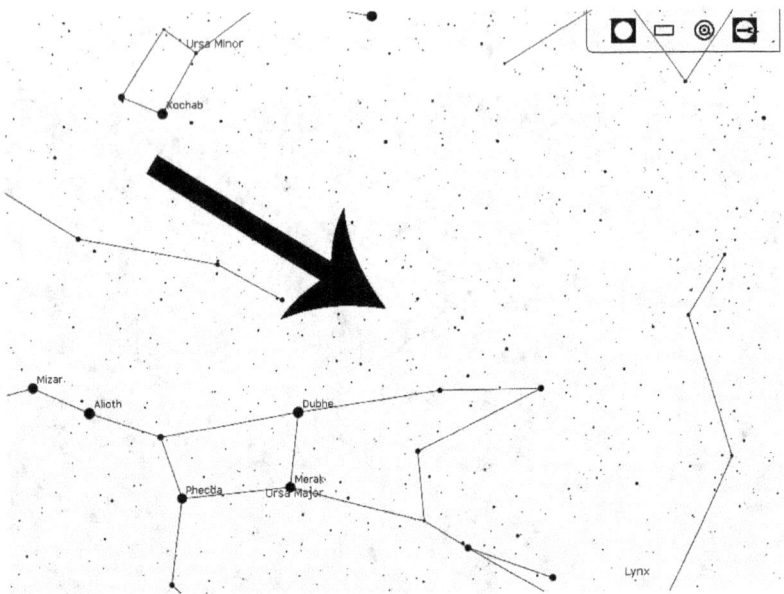

To find these two galaxies simply find Ursa Major, then find Dubhe on his back and the next star towards his nose, 23 Uma. Now look "above" Ursa Major between those two stars about the distance of where Lam Dra is and you should see it.

## #24: The Sculptor Galaxy NGC253

Jan  Feb  Mar  Apr  May  Jun  Jul  Aug  Sep  Oct  Nov  Dec

Point your telescope towards the constellation Sculptor and you can see the Sculptor galaxy. This is a wonderfully bright intermediate spiral galaxy which is easy to see even in binoculars.

While not a face on galaxy, it is at a really nice angle so that you can see the spiral arms (in a moderately sized telescope in reasonably dark skies).

# 50 Amazing Things to See With Your New Telescope

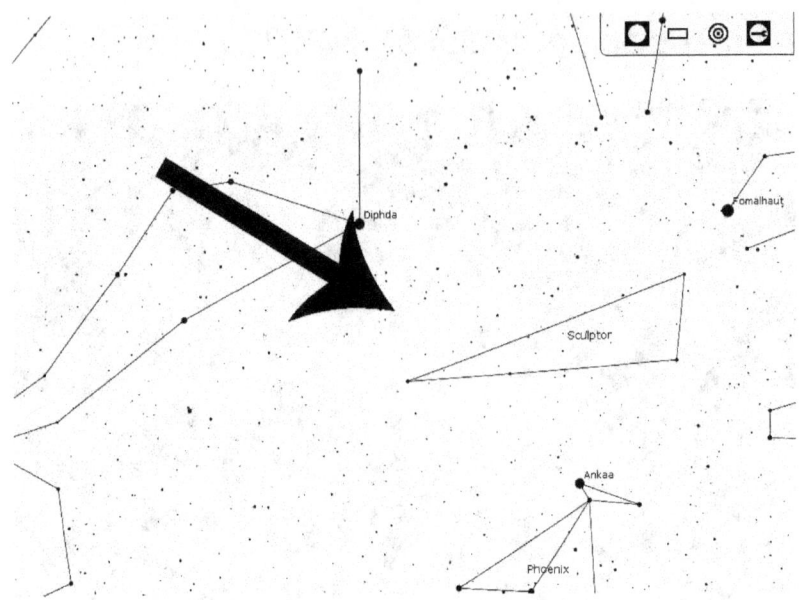

This one is about half way between the star Diphda in Cetus and the star a Scl (alpha Sculptor) in Sculptor.

## #25: Centaurus A NGC5128

Jan  Feb  Mar  Apr  May  Jun  Jul  Aug  Sep  Oct  Nov  Dec

Centaurus A is one of the most unique galaxies you will see in the night sky. Visible only from the southern hemisphere or the very southern portion of the northern hemisphere, it lies in the constellation Centaurus.

This galaxy is so unique, people are still debating what type of galaxy it is. Is it a lenticular galaxy or a giant elliptical? Is it 10 million light years away or 16?

Absolutely worth a look.

# 50 Amazing Things to See With Your New Telescope

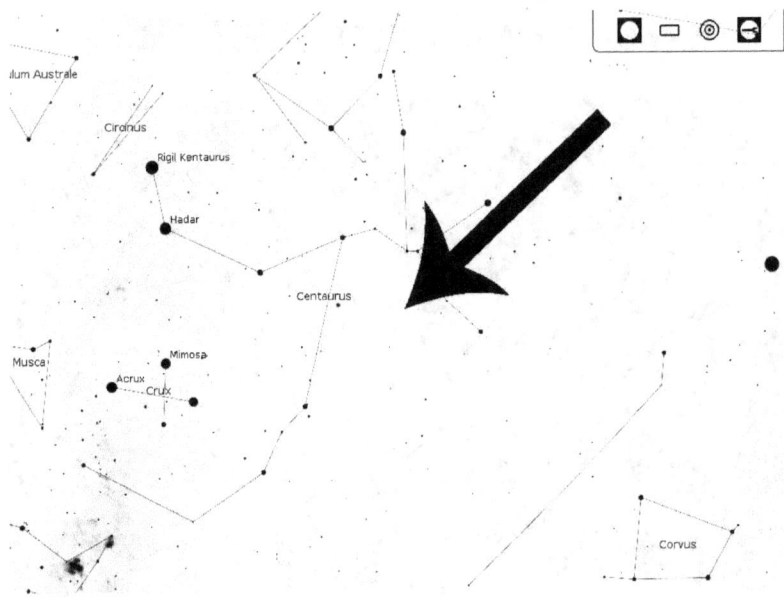

This one is a little more obscure but it is pretty much right behind the right shoulder blade of Centaurus. This puts it surrounded almost equidistantly by stars in the hand, shoulder, back, and rump.

## #26: The Virgo Supercluster

This Virgo Supercluster sounds like a cluster of stars like an open or globular cluster we will talk about later, but it is not. It is a collection of galaxies that you really will not be able to make out too well.

So why in the world would I suggest it as a target to look at if you couldn't make it out very well?

Jan   Feb   Mar   Apr   May   Jun   Jul   Aug   Sep   Oct   Nov   Dec

Because if you look close you can tell there are a few galaxies in your eyepiece, no...wait, looking a little harder there are quite a few galaxies in the eyepiece. Oh, no, there are TONS of galaxies in your eyepiece! How many can you see in the above image?

While the supercluster is a pretty large part of the sky, Markarian's Chain is a section of it that fits into the area of a typical low power eyepiece in a small telescope. This area is a wonderful part of the supercluster to view.

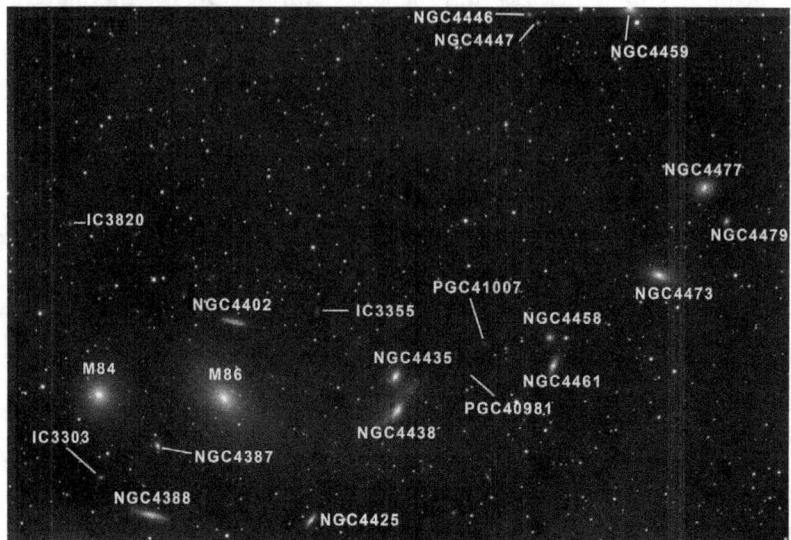

The image above has 21 of the galaxies labeled, but there are still more in the shot. This was captured with a 110mm refractor with a 770mm focal length using a camera that is the equivalent of about a 24mm eyepiece.

While a photograph like the one above certainly shows more detail than a small telescope, a larger telescope would show as much or more detail under dark skies.

# 50 Amazing Things to See With Your New Telescope

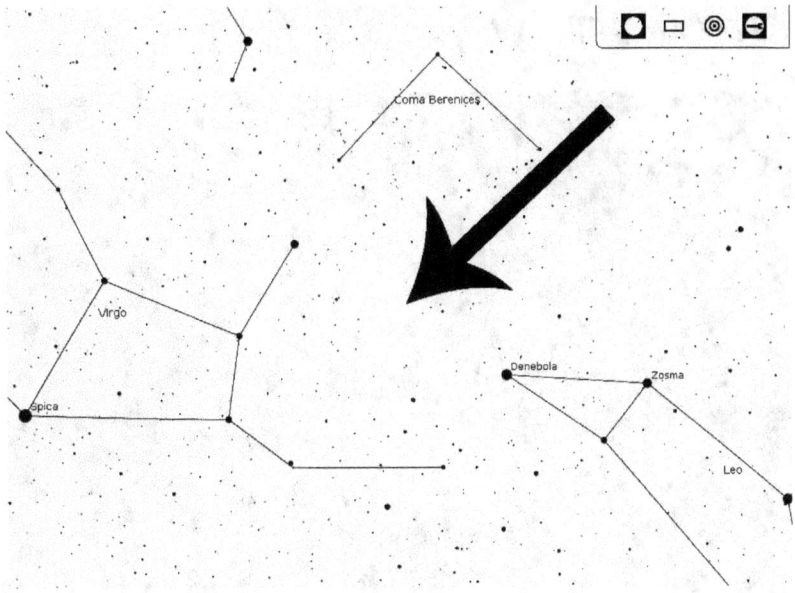

The chain is only part of the supercluster. The chain is located between Coma Berenices, Leo, and Virgo.

# #27: Our Milky Way

Don't forget that our galaxy can be seen from wherever you are, at least a portion of it can be. Here in the middle of the United States I can see it at roughly midnight in June and December high up in the sky.

While you can certainly admire the Milky Way without a telescope or binoculars, it is even more spectacular at low magnification using your telescope.

## 3.4: Star Clusters

Star clusters are areas where stars appear to be in a dense cluster, or in an obvious pattern very close together in the sky.

Clusters come in two basic types; Globular and open.

Globular are clusters that are very densely packed into what might be described as a ball of lights. Think of a string of Christmas lights that are all wadded up in a ball, then plug in that ball of lights and you have a pretty good approximation of a globular cluster.

An open cluster can be that same string of lights after you have untangled it and have it spread out on the floor. They are all still relatively close to each other, and it is still obvious that they are a collection, but they are not all bound up tight either.

In truth, the stars in either type of cluster may not be close to each other at all, they may just appear close because of the way you are looking at them.

Think of two street lights a block away from each other, now move so that they appear to be right next to each other because they are almost both in the same line of sight from your position.

Some clusters have many different types of stars in them, and this can present as many different colors of lights. These can include white, red, yellow, and blue colored stars.

Any way you look at it, clusters are just as beautiful to look at as any Christmas lights.

## #28: The Pleiades M45

Jan   Feb   Mar   Apr   May   Jun   Jul   Aug   Sep   Oct   Nov   Dec

The Pleiades, or Seven Sisters, is one of the most commonly viewed open clusters out there. It is clearly visible from all but the most light polluted skies with just the naked eye.

The name Seven Sisters comes from the fact that someone with very good vision can make out seven distinct stars in the central region of the cluster. These stars appear to be shrouded in a light mist.

You might even notice that the layout of the stars seems a little familiar to you. This could be because it is the basis of the logo on Subaru vehicles.

# 50 Amazing Things to See With Your New Telescope

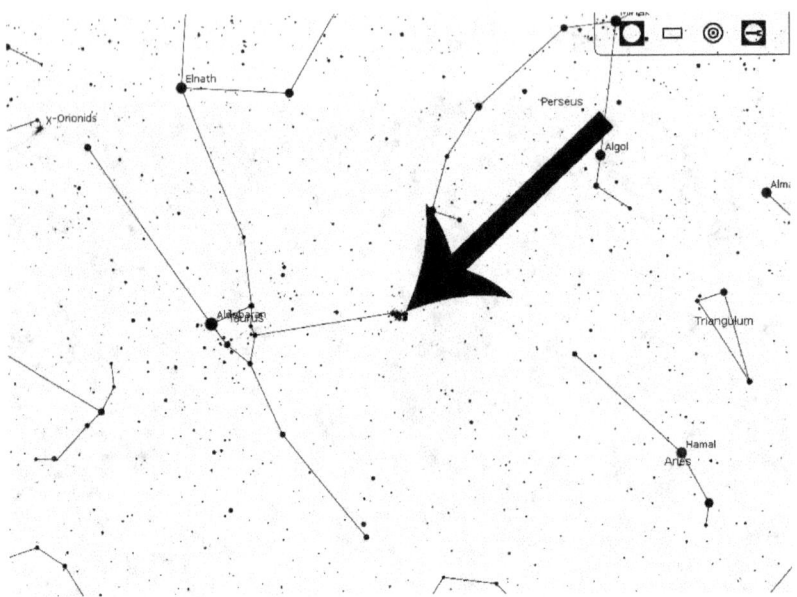

M45 is one of those really bright fuzzy patches in the sky and quite possibly the brightest deep space object. To find it you just find the constellation Taurus and look at the backside of the bull to see the last star, 27 Tau, this is part of M45.

## #29: The Great Globular Cluster in Hercules M13

Jan  Feb  Mar  Apr  May  Jun  Jul  Aug  Sep  Oct  Nov  Dec

If you get to see only one globular cluster, this is the one you want to see. You can clearly see this cluster with the naked eye in moderately dark skies and with a small telescope you should be able to determine what it is.

In medium to larger telescopes you should be able to resolve one of the largest and most beautiful globular clusters in the night sky.

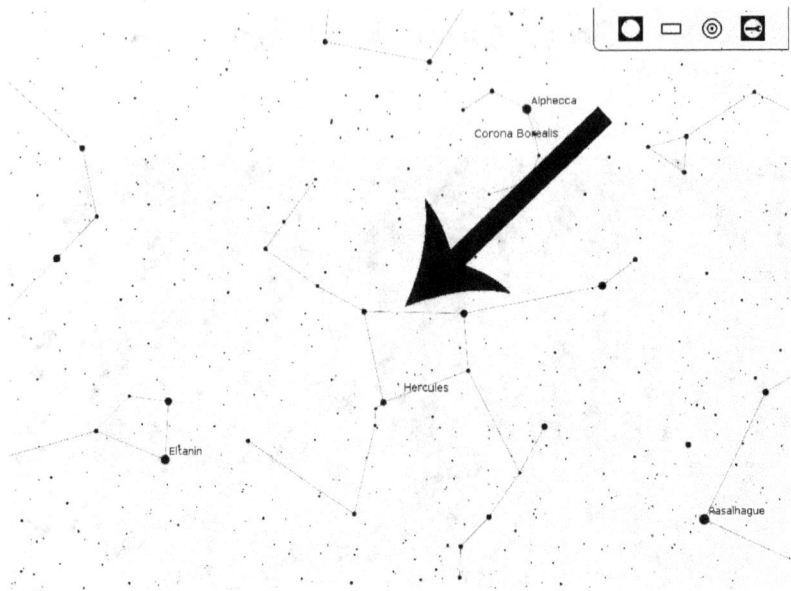

M13 is on the right hip of Hercules in that constellation. It is right on the line of the western side of the box as shown in the above image. A quick scan of the line at low magnification will easily show the cluster.

## #30: The Wild Duck Cluster M11

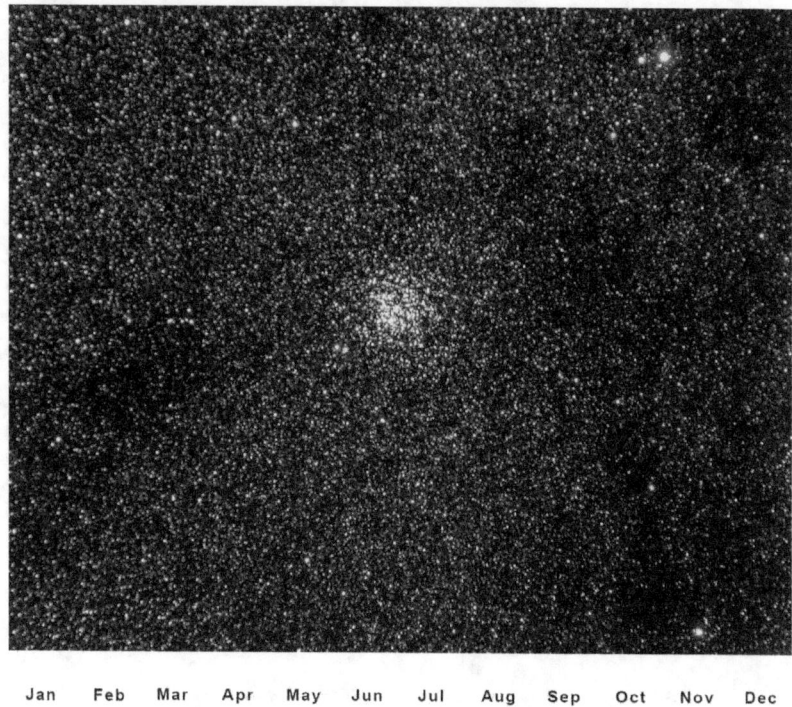

Jan  Feb  Mar  Apr  May  Jun  Jul  Aug  Sep  Oct  Nov  Dec

This open cluster in the constellation of Scutum is so densely packed it could be mistaken for a globular cluster. Thousands of stars are packed in a pretty tight package and can be seen with the naked eye under dark skies.

Once you swing your telescope over you should be able to pick out a ton of stars including many bright blue ones near the center. These are hot new stars just getting their start in life.

This cluster gets its name from the brightest stars which make a sort of triangle shape much like a flock of ducks in flight.

# 50 Amazing Things to See With Your New Telescope

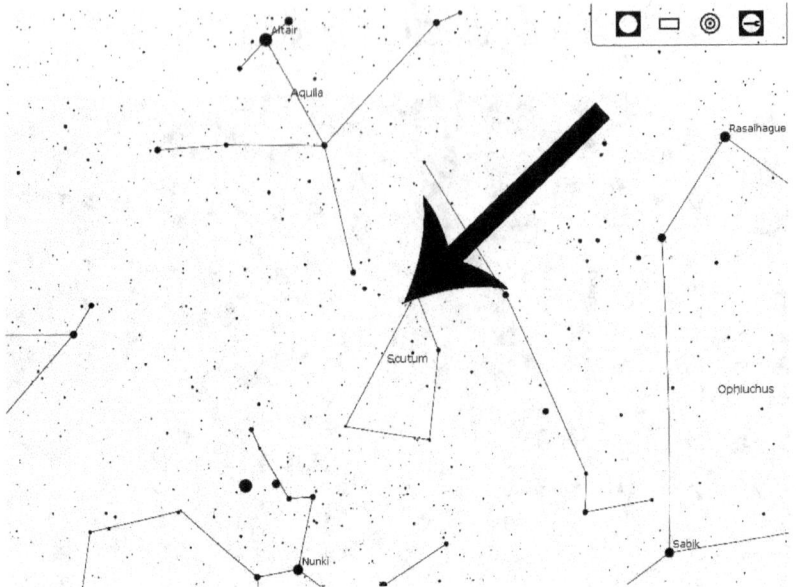

M11 is between Aquila and Scutum on the North West side, closest to Scutum.

## #31: The Beehive Cluster M44

Jan   Feb   Mar   Apr   May   Jun   Jul   Aug   Sep   Oct   Nov   Dec

This open cluster of primarily yellow and blue stars sits in the constellation of Cancer and is a wonderful sight in binoculars or small telescopes.

Larger telescopes do a fine job here as well but tend to show too many stars so that the cluster seems to get swallowed up. I prefer smaller equipment for this target.

# 50 Amazing Things to See With Your New Telescope

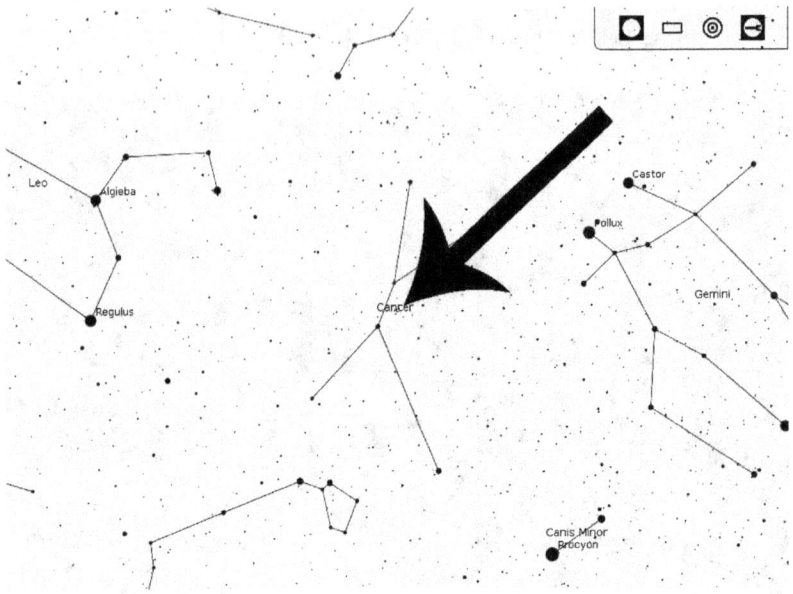

This cluster is almost dead center of the constellation Cancer. In fact if you are looking at the constellation drawings instead of just the diagrams, it would be in the middle of the crab's shell.

## #32: The Scorpius Globular Cluster M4

Jan  Feb  Mar  Apr  May  Jun  Jul  Aug  Sep  Oct  Nov  Dec

Almost the size of a full moon this globular cluster was the first one where stars were resolved to prove its true structure. Even with beginner equipment it should be fairly easy to get a really good look.

With a medium to large telescope the central core will expose a wide variety of colored stars like jewels spread on the sand.

# 50 Amazing Things to See With Your New Telescope

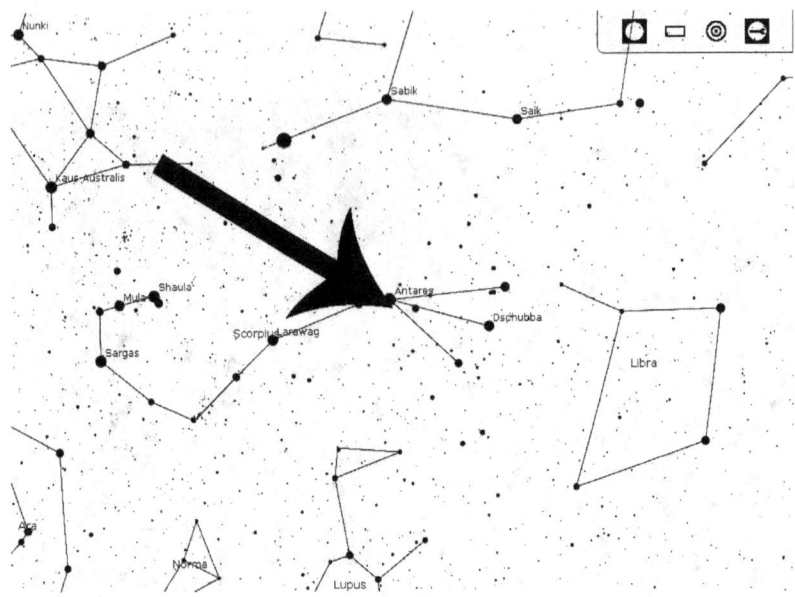

M4 is in Scorpius, just to the south west of Antares right on the line that makes up the right pincer.

## #33: The Pegasus Globular Cluster M15

Jan  Feb  Mar  Apr  May  Jun  Jul  Aug  Sep  Oct  Nov  Dec

Swinging our telescope over towards the constellation of Pegasus we can find one of the oldest known globular clusters, M15. This is another fine example of a large, dense, multicolored, and easy to see globular cluster.

# 50 Amazing Things to See With Your New Telescope

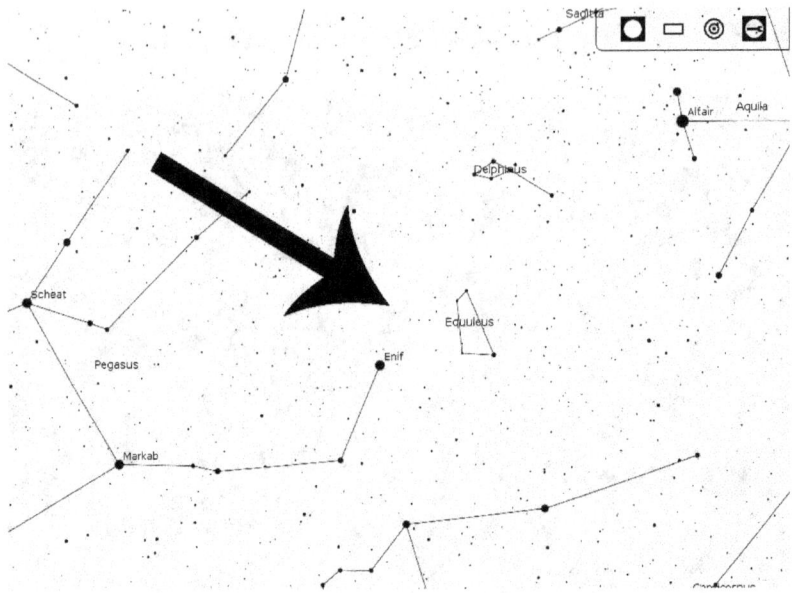

To find M15, find Pegasus and then find the two stars that make up the Pegasus' head. Now extend that line out past Enif (Pegasus' nose) about half the length of the head and you will find the cluster.

## #34: The Double Cluster in Perseus NGC869 & NGC884

Double Cluster in Perseus. Image by Praca własna

Jan  Feb  Mar  Apr  May  Jun  Jul  Aug  Sep  Oct  Nov  Dec

Everyone loves a BOGO (buy one, get one). We even have BOGOs in astronomy as is evidenced by this double cluster.

In between Cassiopeia and Perseus there are two very close and very beautiful open clusters, NGC869 and NGC884. Simply look on the back side of the W formed by Cassiopeia towards Perseus' shoulder and you should easily see them.

# 50 Amazing Things to See With Your New Telescope

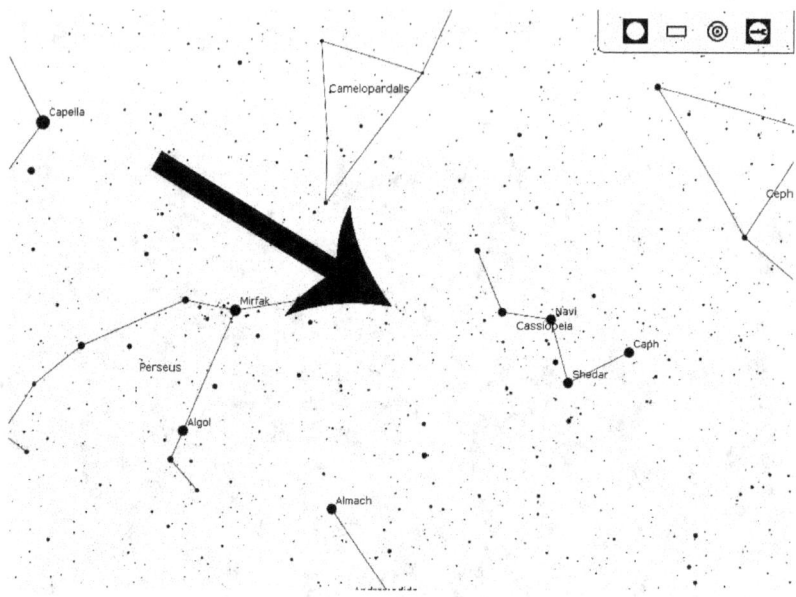

These clusters are right at the top of Perseus' head, between the constellation Perseus and Cassiopeia.

# 50 Amazing Things to See With Your New Telescope

## #35: Omega Centauri NGC5139

Jan  Feb  Mar  Apr  May  Jun  Jul  Aug  Sep  Oct  Nov  Dec

From the northern hemisphere you will need to look pretty low in the sky to the south, but in the southern part of the northern hemisphere (I say this from South Texas) it is absolutely visible.

From anywhere in the southern hemisphere this is a very easy object to find as it is bright and right on the line of constellation of Centaurus. From a moderate to very dark site it can appear almost as large as a full moon.

# 50 Amazing Things to See With Your New Telescope

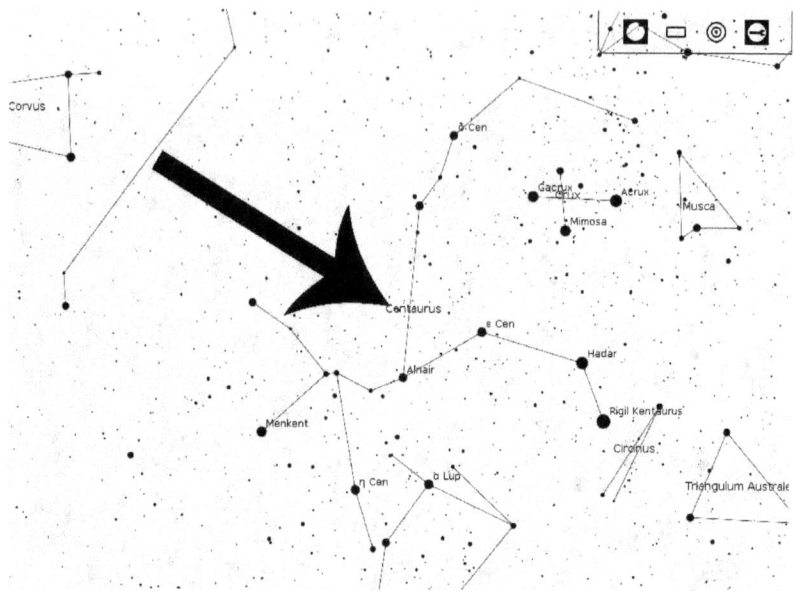

Omega Centauri is right in the middle of the back of the centaur that makes up Centaurus just south west of Zeta Centauri.

## #36: The Jewel Box NGC4755

NGC4755. Image by ESO

Jan  Feb  Mar  Apr  May  Jun  Jul  Aug  Sep  Oct  Nov  Dec

The Jewel Box is an amazing little open cluster in the southern hemisphere in the constellation Crux.

If you are very far south in the northern hemisphere, up on a hill or building top, and have a clear horizon, you could probably see it very low in the sky.

From the southern hemisphere finding it is pretty easy and very rewarding.

# 50 Amazing Things to See With Your New Telescope

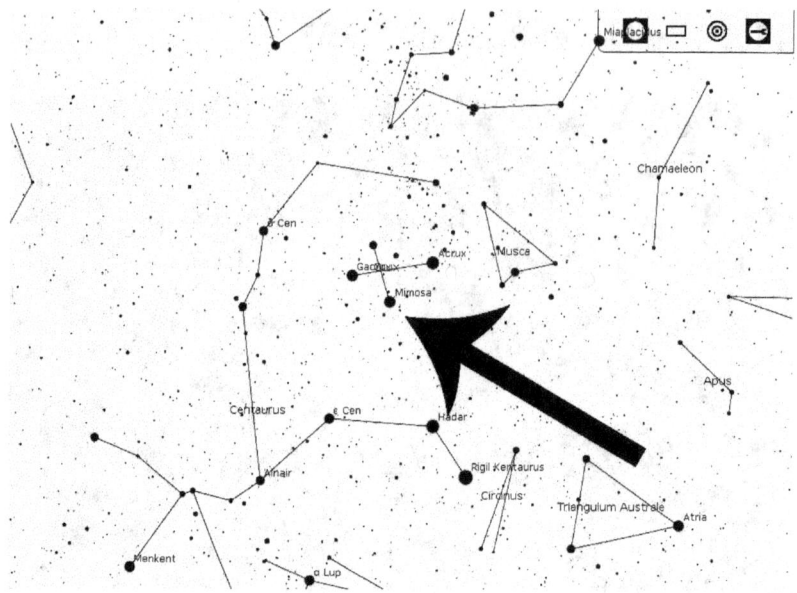

The Jewel Box is located right off the northern side of Crux, beside the star Mimosa.

## #37: 47 Tucanae NGC104

47 Tucanae. Image by: ESO/M.-R. Cioni/VISTA Magellanic Cloud survey.

Jan  Feb  Mar  Apr  May  Jun  Jul  Aug  Sep  Oct  Nov  Dec

This is the second brightest globular in the sky after Omega Centauri and sits in the constellation Tucana. Without being on a skyscraper, you will need to be in the southern hemisphere to see it.

The colors you can see in this cluster are simply amazing from even the city, but especially from a very dark sky.

# 50 Amazing Things to See With Your New Telescope

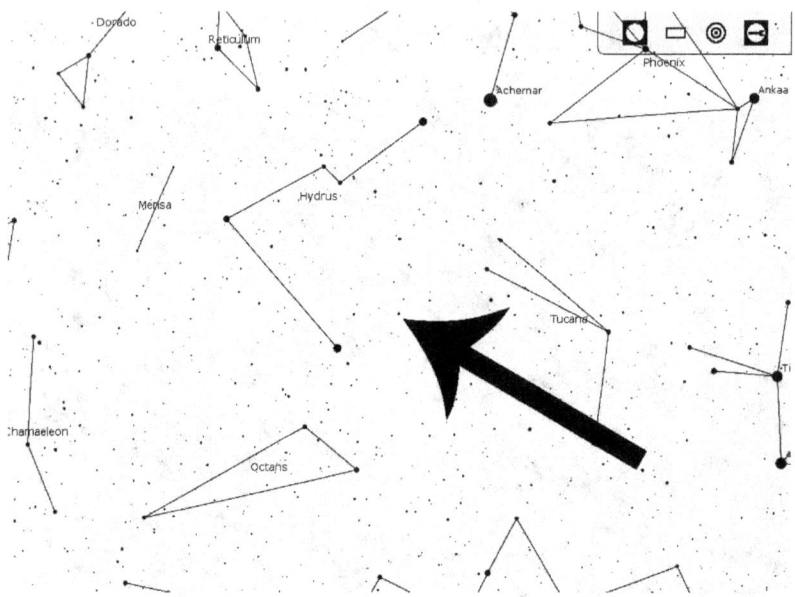

47 Tuc is located between the feet of Tucana and the head of Hydrus. To be a bit more specific, if you drew a line between Zeta Tucana, and Beta Hydrus, right in the middle of that line would be this cluster.

## #38: The Hyades Cluster C41

The Hyades Cluster. Image by: Todd Vance

Jan  Feb  Mar  Apr  May  Jun  Jul  Aug  Sep  Oct  Nov  Dec

This open cluster is almost directly in the middle of the constellation Taurus. One of the most studied open clusters in the sky, it is made up of hundreds of very similar stars roughly in a circle.

The brightest of the stars in the cluster make roughly a V shape which is pretty easy to spot in even a reasonably dark site.

# 50 Amazing Things to See With Your New Telescope

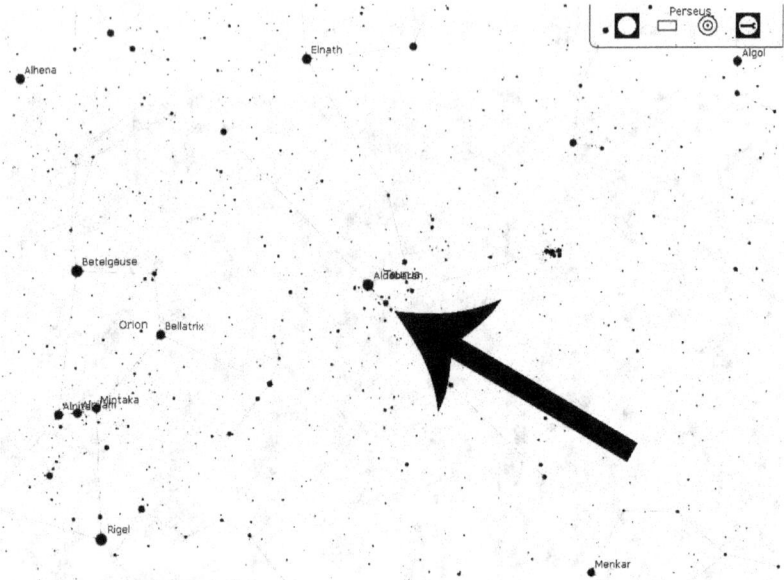

If you can find Taurus, you can find C41 because it is right smack in the middle of the front side of the bull's head.

## 3.5: Stars

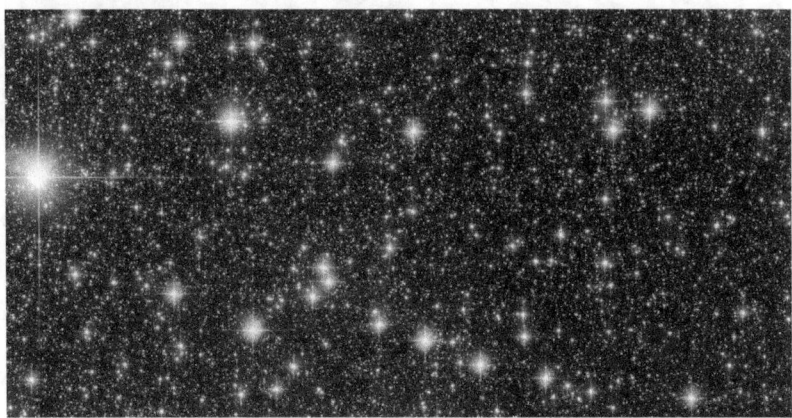

I tend to stay away from recommending you look at individual stars because they just show up as a white dot on a black background. That sounds pretty boring to me.

Even if there is something really special about a particular star, I don't really have any desire to look at an unremarkable white dot on an unremarkable black background.

There are however exceptions to that, and what follows are those exceptions.

## #39: Supernovae

A supernova is when a star explodes at the end of its life. It may sound hard to observe because there is this big kaboom, then it's gone.

Well, yes and no.

A supernova causes the star to blow apart. This could be seen a little too closely as the outer shell of the star expanding outward. Unfortunately you would never really get to see that and live to tell about it.

The good news is that when the star explodes it releases a massive amount of energy and light in that expanding shell. If you are old enough to have ever seen a light bulb get really bright right before it pops and dies, you have an idea of what happens.

The explosion and resulting shell expansion is really bright, and although it is moving really fast, from a great distance you only see the massive brightness. This event can last quite some time so you have plenty of time to read about it and then go out and look for it.

To find supernovae currently visible, look online at https://in-the-sky.org/newsindex.php?feed=novae where they have an excellent list of any and all current events.

## #40: The Double Double (Epsilon Lyrae)

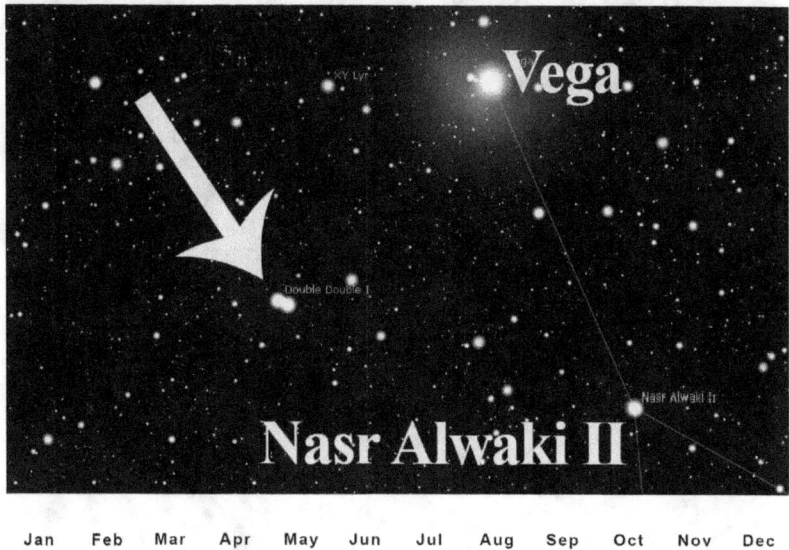

Jan   Feb   Mar   Apr   May   Jun   Jul   Aug   Sep   Oct   Nov   Dec

Looking up at the star Epsilon Lyrae it looks rather unremarkable. Point your binoculars at it and you might be surprised to see that it is actually two stars, or what we call a double (doubles are stars that appear as one until "split" by using enough magnification in a suitable telescope).

Add a little more magnification or a slightly larger telescope and you will find that each of those two stars is actually a double, making this a double double!

# 50 Amazing Things to See With Your New Telescope

This double is easy to split with minimal magnification but you will need a little over 200x to split the double double and see what is shown in the above image.

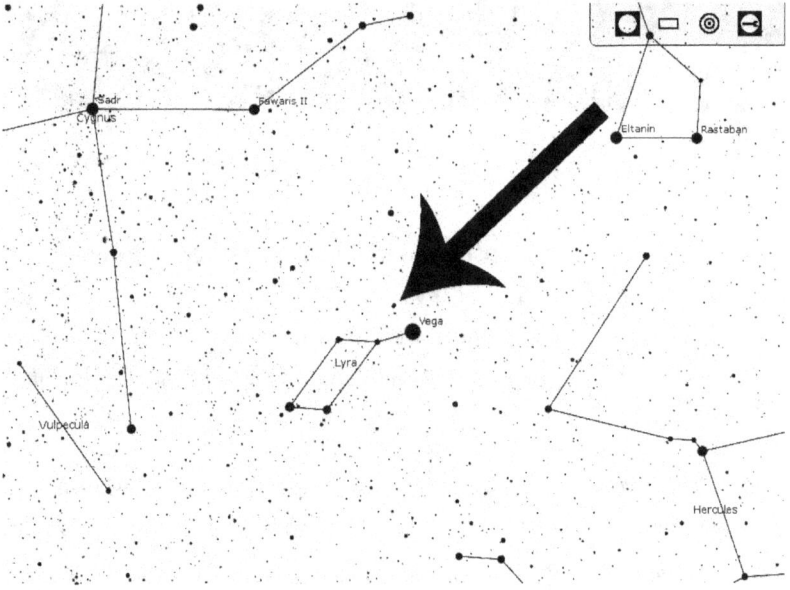

Just off of Vega in the constellation of Lyra.

## #41: The Double Stars of Albireo

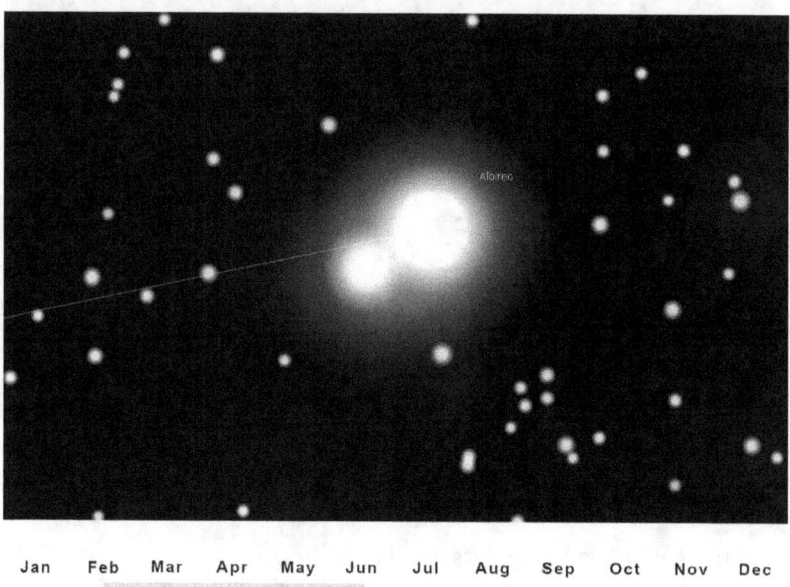

Jan  Feb  Mar  Apr  May  Jun  Jul  Aug  Sep  Oct  Nov  Dec

This is a nice star that is easily split into a double. The primary star (Beta Cygni A) appears yellow while the smaller Beta Cygni B is decidedly blue.

To find it simply find Cygnus, look to the farthest south star in that constellation near Velpecula and that is Albireo.

Since the colors are so dramatically different (not shown in the image above, sorry) it makes for a wonderful view as you split the stars.

# 50 Amazing Things to See With Your New Telescope

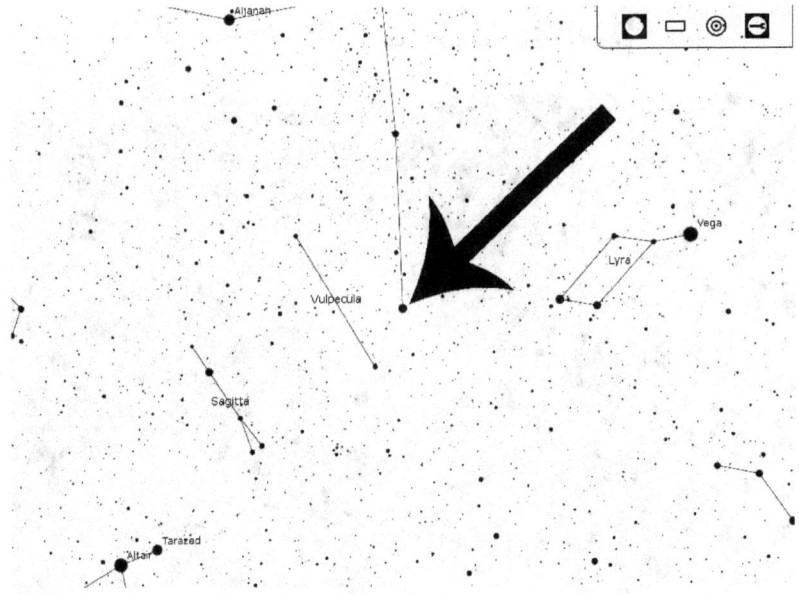

Find the swan Cygnus and its head is Albireo.

## #42: The Double Stars of Almach

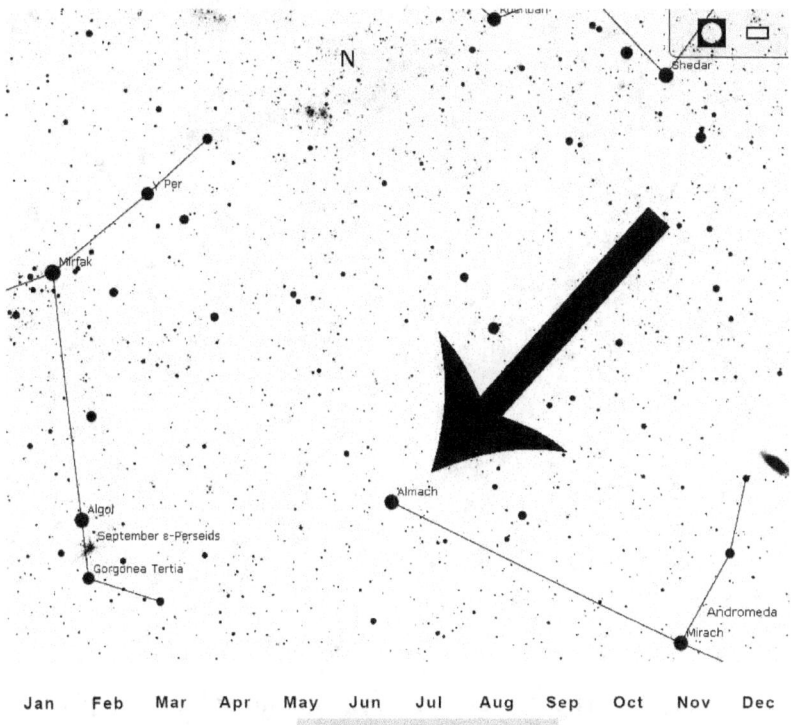

Almach is another fine example of a yellow and blue star pair in the constellation of Andromeda. While not quite as fine a double star as Albireo, it runs a close second.

It is also a bit harder to split so you get more of a challenge.

To find it, look at the absolute northern most star in Andromeda and you will have found Almach.

## #43: The Double Stars of Castor

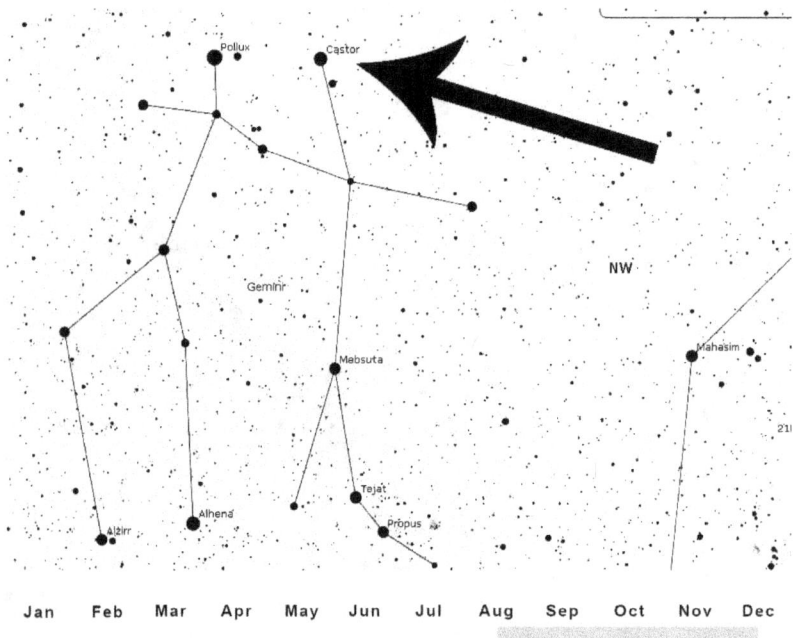

Not as colorful as the previous two, you can still easily split this double in the constellation of the twins with a little over 200x.

It is also easy to find as it is the second brightest star in Gemini and right next to the brightest which is Pollux. Simply find the head of the second twin and you have found Castor.

## 50 Amazing Things to See With Your New Telescope

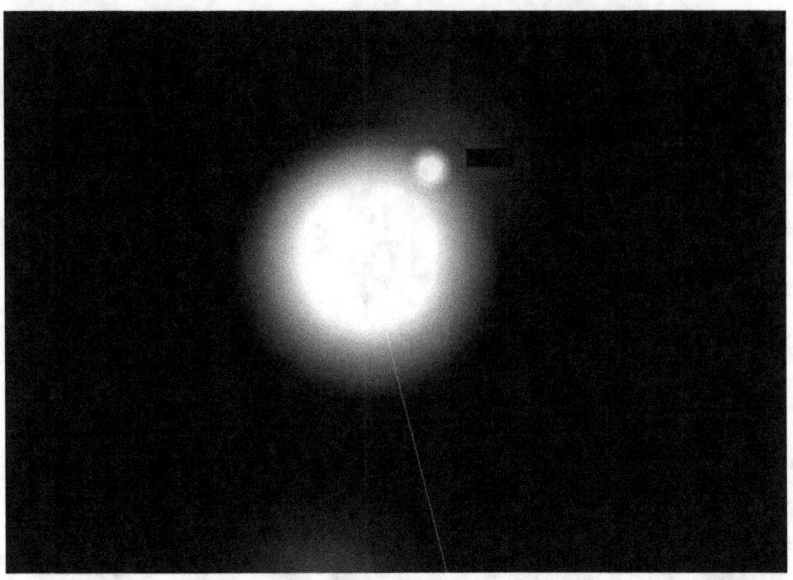

## #44: Constellations

There are 88 modern constellations and some of them will fit into a small telescope eyepiece at low magnification. Those that do not are still a lot of fun to explore one section at a time.

Try navigating from one star to the next in the constellation and you will not only learn to better navigate the night sky, but are sure to find some interesting treats along the way.

The image above shows the belt and sword in Orion which includes the Horsehead nebula, Orion nebula, and De Mairan's nebula among others.

More about the constellations can be found later on in this book.

# #45: Polaris (The North Star)

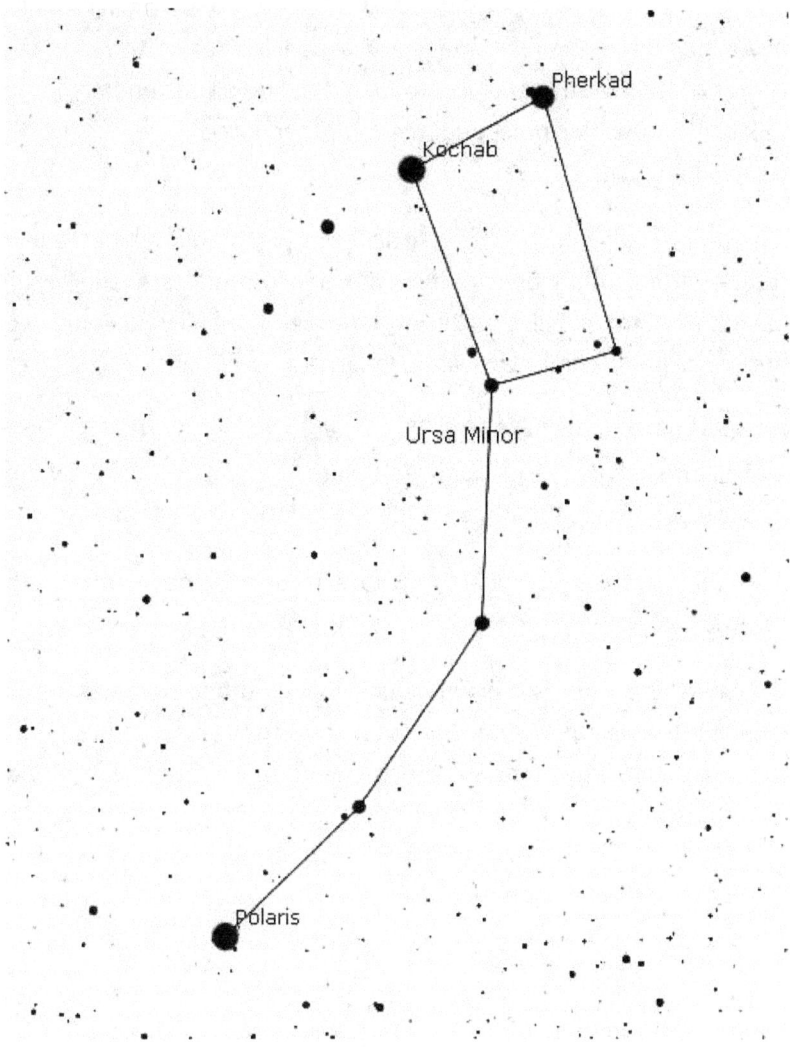

Polaris is a completely unremarkable star. It is not a fancy color, it is not very bright, and it has nothing interesting surrounding it that we know of.

It does have one interesting thing about it, its location.

Polaris is the name for the North Star, and it is, funny enough, pretty much due north in the sky.

If you took a tennis ball and stuck a pencil straight through the center of it so that it stuck out both sides, the eraser end could be the south celestial pole and the point end could be the north celestial pole.

The tennis ball which is our little example of our planet earth rotates around that pencil while the pencil remains fixed. If you looked from the pencil's eraser end through the ball you would see the pencil point pointing towards Polaris.

When standing on the earth, if you were at the North Pole and looked straight up, you would see Polaris. The further towards the equator you move, the lower in the sky Polaris would appear.

Everything in the sky appears to rotate around the north and south celestial poles.

So while the star is unremarkable in all other respects, it is important to be able to identify its location so that you can use that to find other things in the sky (not to mention you can use it to find out where you are and where you need to go).

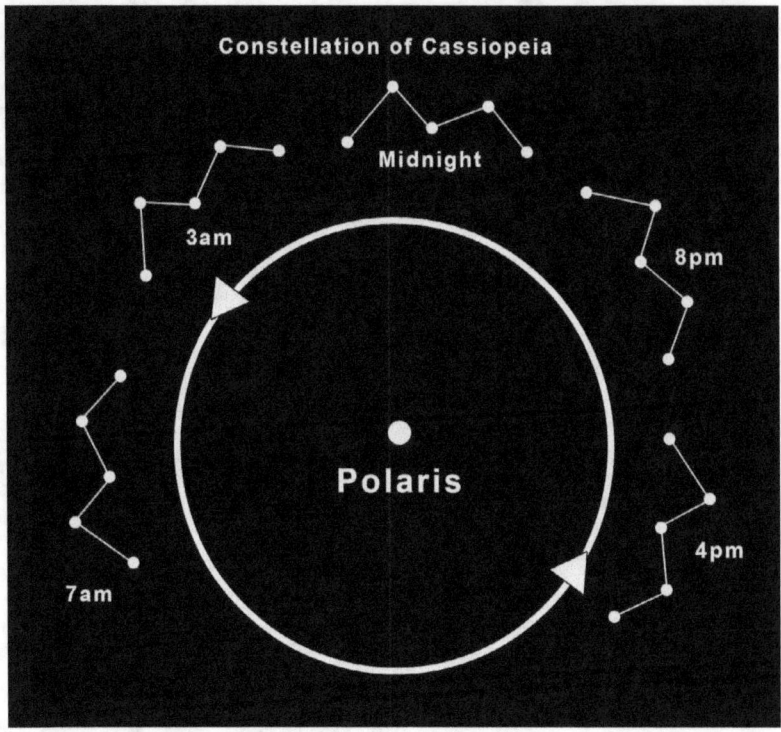

I have a rather unusual way of finding Polaris, I use Cassiopeia instead of finding Ursa Minor. Fortunately I am far enough north that I can always pull this one off but if you are any further south you may not be able to.

# #46: The Southern Cross

There is no southern pole star like Polaris is in the northern hemisphere, but there is a way to find the location of where it would be if there was one, and that is to use the Southern Cross.

And I will bet you thought the Southern Cross was just the name of a song.

While there is no real pole star, Sigma Octantis is the closest at about one degree away. Not only is it a little far off, but it is also pretty dim at magnitude 5.45 (as opposed to 1.98 for Polaris).

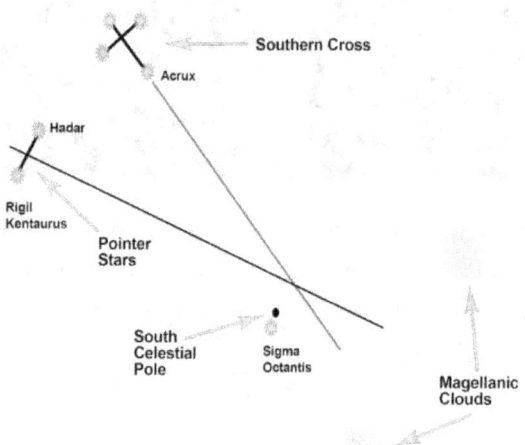

Using the Southern Cross as shown above you can find the southern celestial pole location right next to Sigma Octantis.

# #47: Stellar Spectroscopy

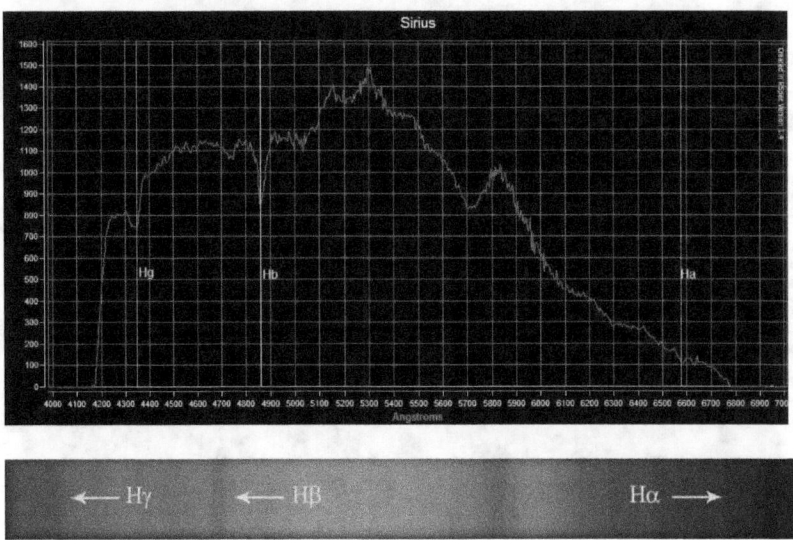

Stellar Spectroscopy is the science of using the light emitted by a star to tell us what that star is made of.

The way we do this is by using a special device that breaks the light up into little slices. Each of these slices represents a particular color, and therefore a particular wavelength, of that light. Different elements emit light at different wavelengths so by using this we can see what elements are present.

While there are very specialized and expensive devices out there for this, we would want to use a very inexpensive and easy method using what is called a grating.

The grating is simply a filter that screws on to the end of our eyepiece and splits out the light for us to view.

This is a fun little specialty which visually you can compare different stars through your telescope to see if you can spot any difference.

For more serious work however you will need a camera. Fortunately, a cheap webcam or DSLR will work just fine. With the camera you can get quite detailed graphs of the spectra of the stars which you can compare to professional results to see how you did. It is a lot of fun!

The images on the previous page were the processed results of the original pictures taken with the filter. Wonder what the original picture looks like?

The above image is a reasonable example of what you might see in the viewfinder. This is the spectra of Arcturus in the constellation Boötes. Keep in mind that this might be printed in a black and white book so you may see nothing but gray whereas in the telescope you will see a range of colors starting with blue on the left, then green and yellow, then finally red.

## 3.6: Events

While a lot of objects in the sky can be viewed for at least several months each year there are some things that happen at a specific time and date. If you miss this event, you miss it.

These events can be some of the most exciting and spectacular things you get to see. Many people (myself included) plan trips to see things like a solar eclipse or the transit of Venus over a year in advance to ensure hotel rooms where we want them.

My last trip to see a total solar eclipse was booked well over a year in advance and things I took into consideration were making my hotel near the optimum viewing city, but still an hour away near the crossroads of major highways.

This was done so that if the weather was poor at the primary location, I would have easy access to major highways moving east and west so that I could pick another location that had better weather.

We spend so much time and effort in seeing these events because some of them may not come again in our lifetime. A good example is the last transit of Venus I saw.

If you missed the transit of Venus, unless we as a species develop robot bodies that we can transfer your consciousness into, or you are one of the oldest people to have ever lived, you will not be alive to see the next one in 2117.

# #48: Conjunctions, Occultations, Transits

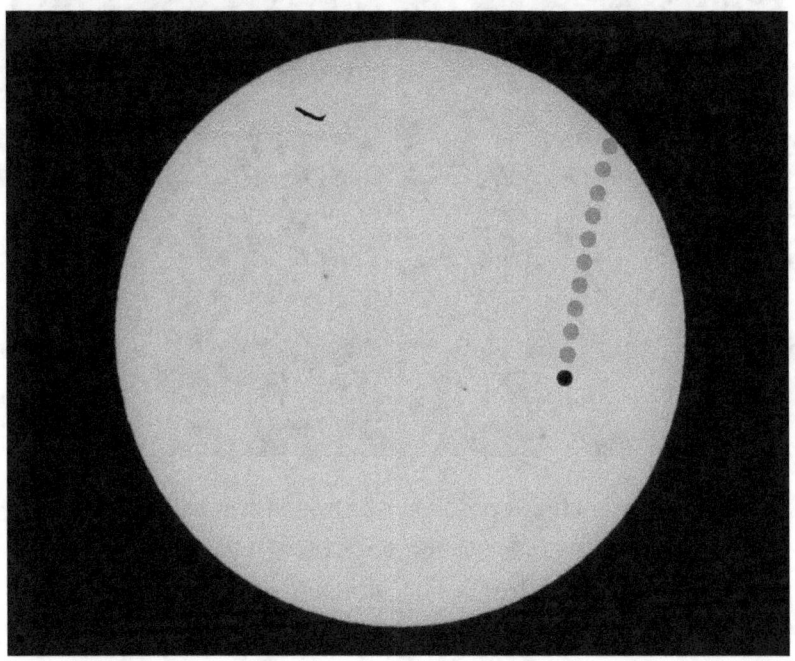

As with a lot of events, the items in question may not be the real treat, it may be where that particular object is in relation to other objects around it.

Conjunctions are a great illustration of this point because they are where multiple objects appear to be in close proximity. A great example is shown on the next page where the moon, Venus, and Mercury all appear in the same widefield shot.

Occultations are when one object passes in front of another and that results in the object in the back being hidden from view. A frequent object in the front of occultations is the moon because of its relatively large size. The moon frequently passes in front of many stars, nebulae, and even planets.

Transits are pretty much the opposite of occultations in that we are concerned with the object passing in front. Common transits of importance include the Venus and Mercury transits of the sun.

## #49: Solar Eclipses

A solar eclipse is where the moon passes directly between the earth and the sun, preventing some of the sunlight from striking the earth. There are three different types of solar eclipses; total, annular, and partial.

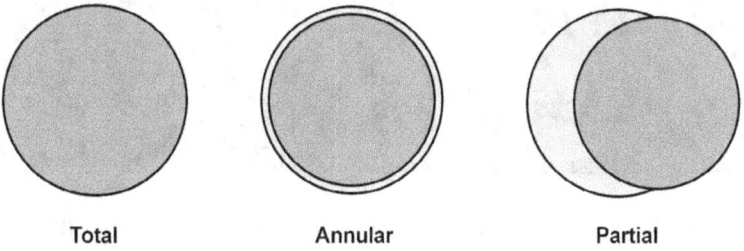

Total        Annular        Partial

A total solar eclipse is exactly what it sounds like, the moon totally covers the sun and you can look directly at the eclipse (when it is in totality, the phase where the moon completely covers the sun) with the unaided and unprotected eye. At all points leading up to and after the totality of the eclipse you will need professionally made (though inexpensive) eye or telescope protection.

If you have never seen a total solar eclipse and you ever get the chance, go for it. I have witnessed a lot of celestial events in my life, none of which compare to a total solar eclipse.

You might think it is simply a matter of it getting dark in the middle of the day, and then light again. Maybe a pretty cool sight when you get to look right at it. That is a massive understatement and oversimplification of what happens.

The lighting leading up to and during the eclipse is like nothing you have ever seen. It is also like nothing the animals and insects have ever seen and they get confused too. This results in everything looking, acting and sounding like it came right out of a science fiction movie.

Trust me, if you have never watched a total solar eclipse, you have absolutely no idea what you are missing.

May 20th, 2012 Annular Eclipse - Albuquerque, New Mexico - Allan Hall

An annular solar eclipse is also a very cool thing to participate in. While the sun is not completely covered, it covers the inside of the sun leaving a "ring of fire" around the outside edge of the moon. This is also referred to an annulus, hence the term annular eclipse.

Annular eclipses happen more often than total solar eclipses so it should be easier for you to catch one. Even in the middle of this type of eclipse when the moon is blocking most of the sun, you will need professionally made protection for your telescope or your eyes to look at the eclipse.

The last type of solar eclipse is the partial, and that is exactly what it sounds like. Partial eclipses can be where the sun is mostly blocked and the sky get very dark, to something where the moon blocks so little of the sun that without a telescope and solar filter you would never even know it happened.

Even a small partial eclipse can be a lot of fun to view for you and your family or group of astronomy buffs. Be sure to take a lot of sunblock lotion, and drink lots of water!

To find the date of the next solar eclipse in your area you can visit allans-stuff.com/astronomy-books/astronomical-events/ for many upcoming dates.

# #50: Lunar Eclipses

A lunar eclipse is where the earth comes between the sun and moon, blocking the sunlight from reaching the moon. This is most apparent in a total lunar eclipse where the moon is perfectly behind the earth.

One major difference is that the earth cannot block 100% of the sunlight which means that a little light still reaches the surface of the moon. This light travels through the atmosphere on the outside edges of the earth and therefore (just like a sunrise or sunset) turns that light red.

This effect causes the moon's surface to substantially darken and to turn reddish, but does not cause the moon to turn completely dark like it does during a new moon.

Similar to a solar eclipse there are different types of lunar eclipses; total and partial.

The total is what we have already discussed whereas the partial allows some direct sunlight to strike the surface. A partial eclipse could present as just a very slight darkening of the moon's surface all the way through to what looks very much like a total eclipse.

Lunar eclipses are more common than solar eclipses so finding one you can view should not be a problem. To view upcoming eclipses you can access the list at allans-stuff.com/astronomy-books/astronomical-events/ which includes both solar and lunar eclipse dates.

50 Amazing Things to See With Your New Telescope

## 4: The Constellations

There are 88 constellations in the night sky and they are:

| | | |
|---|---|---|
| Andromeda | Cygnus | Pavo |
| Antlia | Delphinus | Pegasus |
| Apus | Dorado | Perseus |
| Aquarius | Draco | Phoenix |
| Aquila | Equuleus | Pictor |
| Ara | Eridanus | Pisces |
| Aries | Fornax | Piscis Austrinis |
| Auriga | Gemini | Puppis |
| Boötes | Grus | Pyxis |
| Caelum | Hercules | Reticulum |
| Camelopardus | Horologium | Sagitta |
| Cancer | Hydra | Sagittarius |
| Canes Venatici | Hydrus | Scorpius |
| Canis Major | Indus | Sculptor |
| Canis Minor | Lacerta | Scutum |
| Capricornus | Leo | Serpens |
| Carina | Leo Minor | Sextans |
| Cassiopeia | Lepus | Taurus |
| Centaurus | Libra | Telescopium |
| Cephus | Lupus | Triangulum |
| Cetus | Lynx | Triangulum Australe |
| Chamaeleon | Lyra | Tucana |
| Circinus | Mensa | Ursa Major |
| Columba | Microscopium | Ursa Minor |
| Coma Berenices | Monoceros | Vela |
| Corona Australis | Musca | Virgo |
| Corona Borealis | Norma | Volans |
| Corvus | Octans | Vulpecula |
| Crater | Ophiuchus | |
| Crux | Orion | |

# 50 Amazing Things to See With Your New Telescope

## 4.1: Constellation Charts

The next four charts show where the constellations are overhead at a specific time of the night on a specific date over the central US. These are meant to give you an idea of where the constellations are in relation to one another, and how they move from season to season.

Your planisphere or https://in-the-sky.org/ will be able to give you more accurate maps for your location, date, and time.

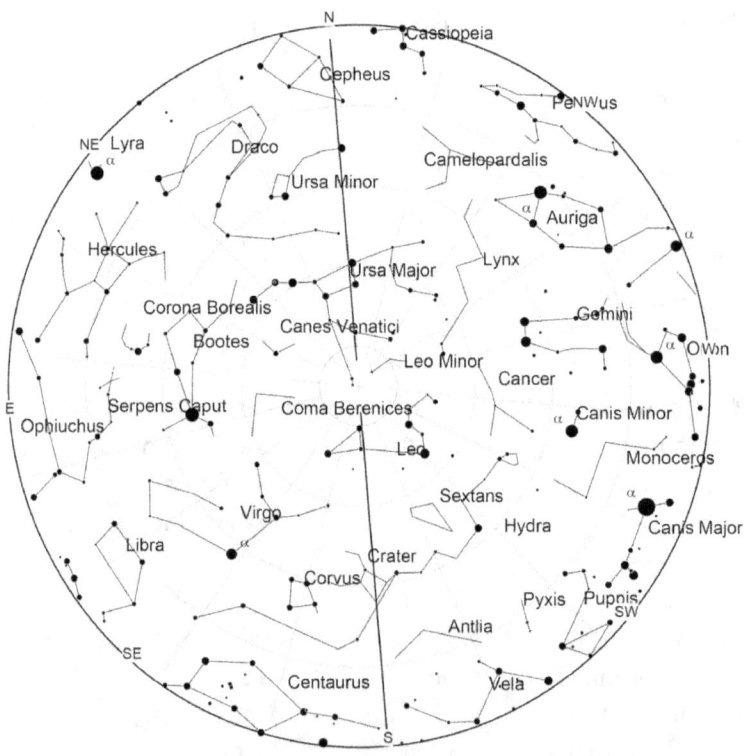

Spring sky, south central US, May 15th 9pm CST.

Galaxies galore as Virgo creeps to center stage.

# 50 Amazing Things to See With Your New Telescope

Summer sky, south central US, August 15th 9pm CST.

Messier 13 is the star of the show with Hercules overhead.

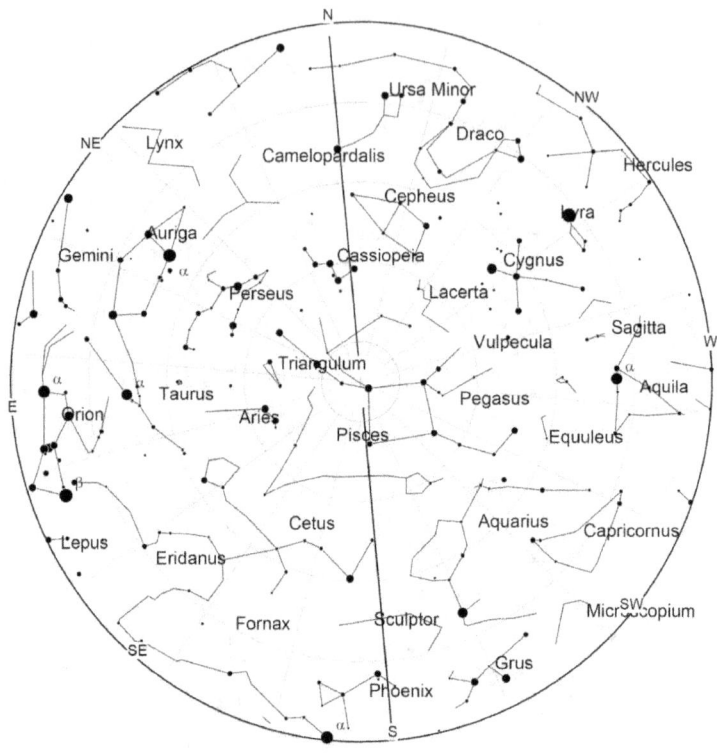

Fall sky, south central US, November 15th 9pm CST.

Andromeda is high in the sky as we get our last glimpse of Hercules.

# 50 Amazing Things to See With Your New Telescope

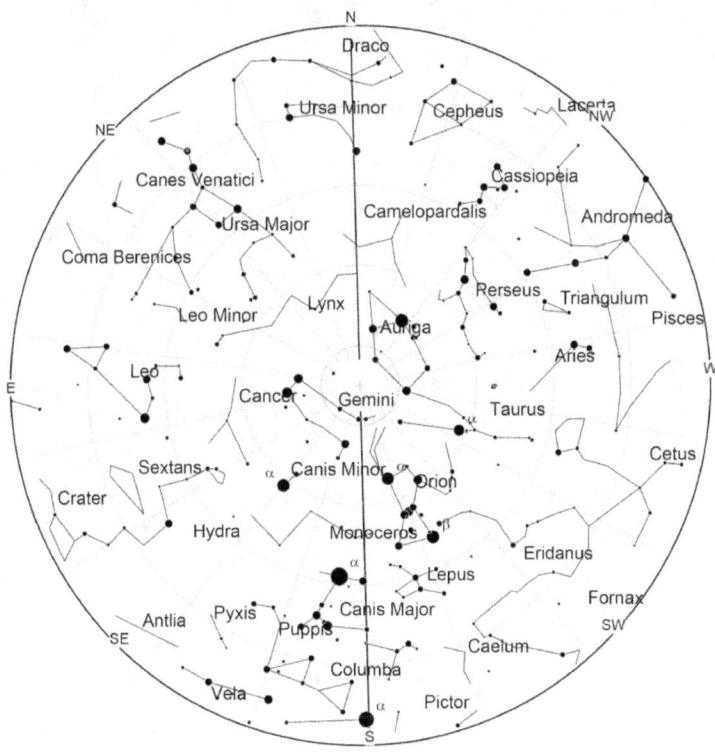

Winter sky, south central US, February 15th 9pm CST.

The nights are getting cold while M45 rises early leading Orion the Hunter across the sky. Get there early and see Andromeda and her galaxy before she descends below the horizon. This is a favorite time of the year for many astronomers.

# 5: Where to go from here

Oh boy, are there a lot of places you can go so here are some suggestions:

**Astronomy equipment:**
| | |
|---|---|
| Orion Telescopes | -www.telescope.com |
| Agena Astro | -www.agenaastro.com |
| Oceanside Telescope | -www.optcorp.com |
| Astromart (used) | -www.astromart.com |
| ScopeStuff | -www.scopestuff.com |

**Online forums:**
| | |
|---|---|
| Astronomy Magazine | -www.astronomy.com |
| Stargazers Lounge | -www.stargazerslounge.com |
| Cloudy Nights | -www.cloudynights.com |
| Ice In Space | -www.iceinspace.com |
| Astromart | -www.astromart.com/forums/ |
| Telescope Junkies | -www.telescopejunkies.com |

**Specializations:**
| | |
|---|---|
| Spectroscopy | -www.rspec-astro.com |
| Radio Astronomy | -www.radio-astronomy.com |
| Photometry | -www.citizensky.org |
| Astrophotography | -www.allans-stuff.com |

**Planetarium software:**
| | |
|---|---|
| TheSkyX | -www.bisque.com |
| Starry Night | -www.starrynight.com |
| Stellarium | -www.stellarium.org |
| Cartes du Ciel | -www.ap-i.net |
| C2A | -www.astrosurf.com |

**Session planning software:**
| | |
|---|---|
| Astroplanner | -www.astroplanner.net |
| Skytools | -www.skyhound.com |
| Deep Sky Planner - | -www.knightware.biz |

## 5.1: Best months to view objects in this book

| Jan | Mar | Apr | May | Jun | Jul | Aug | Oct | Nov | Dec |
|---|---|---|---|---|---|---|---|---|---|
| M81 | Epsilon Lyrae | Virgo SC | M51 | NGC253 | M8 | M27 | M31 | M45 | M42 |
| M82 | | M84 | M63 | Milky Way | M16 | M15 | M33 | | M1 |
| M44 | | M86 | | M13 | M17 | C41 | Castor | | NGC3242 |
| | | Albireo | | M4 | M20 | | | | NGC5128 |
| | | | | NGC104 | M57 | | | | Milky Way |
| | | | | | M11 | | | | NGC5139 |
| | | | | | NGC869 | | | | NGC4755 |
| | | | | | NGC884 | | | | |
| | | | | | Almach | | | | |

## 5.2: Index

23 Uma, 64
27 Tau, 76
**47 Tucanae**, 93
a Scl, 66
**Albireo**, 103, 104
Alkaid, 60, 62
**Almach**, 105
Andromeda, 8, 55, 56, 58, 105
**Andromeda Galaxy**, 55
Antares, 84
**Asteroids**, 24
**Beehive Cluster**, 81
Beta Hydrus, 94
**Bode and Cigar Galaxies**, 63
C2A, 131
**C41**, 95
Caldwell, 7
Cancer, 81, 82
Canes Venatici, 59, 60, 62
Cassiopeia, 87, 88, 111
**Castor**, 106
catalog number, 7
catalogs, 7
Centaurus, 8, 67, 68, 89, 90
**Centaurus A**, 67
**Ceres**, 24
Cetus, 66
Coma Berenices, 71
**Comets**, 31
**Conjunctions**, 117
**Constellations**, 108
Cor Caroli, 60, 62
**Crab Nebula**, 49
Crux, 91, 92

Cygnus, 103, 104
Delta Aquarids, 29
Diphda, 66
**Double Cluster in Perseus**, 87
**Double Double**, 101
Draconids, 29
Dubhe, 64
**Dumbbell Nebula**, 43
**Eagle Nebula**, 39
Enif, 86
**Epsilon Lyrae**, 101
Eta Aquarids, 29
**Events**, 115
**Galaxies**, 53
Geminids, 29
**Ghost of Jupiter**, 51
**Great Globular Cluster in Hercules**, 77
Hercules, 8, 78
**Hyades Cluster**, 95
Hydra, 52
Hydrus, 94
**International Space Station**, 27
**Iridium Flares**, 25
**Jewel Box**, 91
**Jupiter**, 22
**Lagoon Nebula**, 37
Lam Dra, 64
Leo, 71
Leonids, 29
**Lunar Eclipse**, 122
Lyra, 29, 48, 102
Lyrids, 29
**M1**, 49

133

**M11**, 79
**M13**, 77
**M15**, 85
**M16**, 39
**M17**, 41
**M20**, 45
**M27**, 43
**M31**, 55
**M33**, 57
**M4**, 83
**M42**, 35
**M44**, 81
**M45**, 75
**M51**, 59
**M57**, 47
**M63**, 61
**M8**, 37
**M81**, 63
**M82**, 63
Markarian's Chain, 69
Messier, 7
**Meteor Showers**, 28
meteorite, 28
Meteoroids, 28
Meteors, 28
Mimosa, 92
**Moon**, 13
moon filter, 15
Mu Hydrae, 52
**Nebulae**, 33
New General Catalog, 7
**NGC104**, 93
**NGC253**, 65
**NGC3242**, 51
**NGC4755**, 91
**NGC5128**, 67
**NGC5139**, 89

**NGC869**, 87
**NGC884**, 87
**North Star**, 109
Nu And, 56
**occultations**, 117
**Omega Centauri**, 89
**Omega Nebula**, 41
**Orion Nebula**, 35
Orionids, 29
Pegasus, 8, 85, 86
**Pegasus Globular Cluster**, 85
Perseids, 29
Perseus, 8, 87, 88
Pisces, 58
Planisphere, 5
**Pleiades**, 75
**Polaris**, 109
Polis, 46
Pollux, 106
Quadrantids, 29
**Ring Nebula**, 47
Sagitta, 44
Sagittarius, 37, 38, 42, 46
**Saturn**, 20
Scorpius, 8, 83, 84
**Scorpius Globular Cluster**, 83
Sculptor, 8, 65, 66
**Sculptor Galaxy**, 65
Scutum, 40, 42, 79, 80
Sheliak, 48
Sigma Octantis, 112
**Solar Eclipse**, 119
<u>solar filter</u>, 17
solar finder, 19
**Southern Cross**, 112
**Star Clusters**, 73
**Stars**, 97

**Stellar Spectroscopy**, 113
Sulafat, 48
**Sun**, 17
**Sunflower Galaxy**, 61
sunspots, 18
**Supernovae**, 99
Taurids, 29
Taurus, 49, 50, 76, 95, 96
**transits**, 117
Triangulum, 8, 57, 58
**Triangulum Galaxy**, 57
**Trifid Nebula**, 45

Tucana, 93, 94
Ursa Major, 60, 62, 64
Ursids, 29
**Vesta**, 24, 46
Virgo, 8, 69, 71
**Virgo Supercluster**, 69
Vulpecula, 44
**Whirlpool Galaxy**, 59
**Wild Duck Cluster**, 79
Zeta Centauri, 90
Zeta Tucana, 94

## 5.3: Glossary

**A/D converter (ADC)** - Analog to digital converter. A camera sensor records light as an analog signal which the A/D converter then converts into digital information.

**Achromat** – A type of refractor typically with two lens elements to correct for chromatic aberrations. This type of scope is not well suited for astrophotography.

**Afocal** - A means of taking an image through an eyepiece of a telescope without removing the lens from the camera.

**Alt/Az** - Altitude Azimuth, a type of telescope mount that moves up and down, left and right as opposed to the smooth rolling motion of an EQ mount which accurately tracks the motion of the stars around the earth.

**Amp glow** – Amp glow is the glow that some cameras show on a long exposure image. This usually manifests itself in the corners of the image first and then can spread towards the center. A moderate amount of this can be removed using dark frames. Severe cases cannot be corrected.

**Aperture** - In telescopes, the diameter of the opening at the front of a telescope, usually measured in millimeters. Can also be measured in inches for larger scopes. In camera lenses there is a diaphragm inside the lens that controls the aperture which is sometimes referred to as an F-Stop.

**Apochromatic (APO)** – A type of refractor extremely well adjusted to remove most or all chromatic aberrations which makes it excellent for astrophotography uses. Can have two, three, or more lens elements. Higher end versions almost always have three or more elements.

**Arc Minute** – There are 360 degrees in the sky as it goes 360 degrees around us. One arc minute is $1/60^{th}$ of a degree.

**Arc Second** – Is equal to $1/60^{th}$ of an arc minute.

**Artifacts** - Errors or unwanted signals in the image.

**ASCOM** - abbreviation for AStronomy Common Object Model and is a standard in the astronomy equipment industry for control interface design of astronomical equipment such as mounts, focusers, motorized domes, etc.

**Astrograph** - A type of Newtonian telescope that is designed specifically for astrophotography.

**Astrometry** – Extremely precise measuring of objects like comets and asteroids.

**Astrophotography** - Photography of objects in the sky.

**Autoguider** - A camera and associated equipment used to increase the accuracy of the mount in tracking the stars.

**Audio Video Interleave (AVI)** – A wrapper for computer video files, can contain a variety of different formats, typically video for Windows formats, and has a file extension of .AVI.

**Back Focus** – The necessary distance needed to be able to attach a camera onto a telescope focuser, and be able to bring the image projected onto that camera's sensor into focus.

**Backlash** – Unwanted spacing between gear assemblies usually resulting in some "play" or "slop" with the device. This is normally used to describe issues with a mount but can be applied to anything with gears.

**Baffles** – Ridges running around the inside of the light path in a telescope to prevent the scatter of light inside the telescope and provide an image with greater contrast.

**Bahtinov mask** - A mask or cover that goes in front of a telescope with a specific pattern of slits designed to provide easy focusing of point light sources such as stars.

**Barlow** - An optical device that increases the magnification or reduces the field of view, depending on how you look at it. This trades some image quality and light for more magnification. These plug into the optical train just before the eyepiece.

**Bayer matrix** - In color one shot cameras (any camera that produces a single color image in one exposure) the pixels are grouped in groups of four, one red, one blue and two green. These are combined to generate the color information for that area of the image. The Bayer matrix is the array of colored filters over the pixels that accomplishes this.

**BFA** – Bayer Filter Array, see Bayer matrix above.

**Bias frame** - An image taken with the highest shutter speed possible on a given camera at the same ISO and temperature of the light frames. This is used to subtract the camera's electrical signal present in every frame it takes from the final image.

**Binning** – A process of combining multiple pixels in order to boost sensor sensitivity at the expense of resolution. For example, 1x1 binning means each pixel counts as one pixel and is in effect not binned, 2x2 binning would take a square of 4 pixels and combine them into one "super pixel".

**Binos** - Short for binoculars.

**Bino-Viewer** - A device that allows attaching two eyepieces to a standard telescope so you may view objects in stereo.

**Bit** - A single bit can be either on or off, representing either 0 or 1. Computers use this as the basic language of everything they do.

**Bit depth** - This describes a measurement of something like the number of colors an image can contain and is base two mathematics. An example is a 1 bit scale will contain two possible combinations, a 2 bit scale will contain 4, a 4 bit scale will contain 16 and an 8 bit scale will contain 256 bits.

**Black point** - An area of an image that represents absolute black.

**Blooming** - In a camera, once a pixel has received as much light as it can handle, the voltage can spill over into adjacent pixels causing them to be brighter than they should.

**Bortle scale** – Astronomer John Bortle developed a scale of nine levels which represents the "true darkness" of a site, or the amount of light pollution present.

**Bulb exposure** – A bulb exposure is an exposure where as long as the shutter button is held the camera continues the exposure. DSLRs and other cameras can be used in this mode.

**CCD** - Short for Charged-Coupled Device, a type of sensor used in digital cameras. In astrophotography it is usually used as a reference to a camera designed and used specifically for astrophotography as opposed to a digital SLR or other multi use digital camera.

**Celestial equator** - An imaginary line which is basically the equator of the earth projected up into the sky.

**Center mark** – A dot placed exactly in the center of the primary mirror of a Newtonian to aid in collimation.

**Chromatic aberration** – Chromatic aberration is the "glowing" or "fringing" of light around bright objects in a telescope. This is caused when light passes through the optical path it is split into its component colors and then rejoined imperfectly at the focal point.

**Clip** - Clipping an image means you have cut off one end or the other of the image's ability to record data (as can be shown in a histogram). Clipping the highlights for example means that area of the image is pure white and cannot contain any detail. Clipping the darks means that part of the image is pure black and contains no detail.

**CMOS** - Complimentary Metal Oxide Semiconductor. In astrophotography, a type of sensor in a camera.

**Collimation** - The act of aligning the optical components of a telescope to make sure all parts of an image combine correctly into one sharp image.

**Coma** - An optical defect normally present in reflector telescopes that can cause point light sources such as stars to appear to be out of round, presenting like they have the tail of a comet.

**Coma corrector** - An optical device for reflector telescopes to correct for coma aberrations.

**Convolution** – A mathematical method of multiplying arrays of numbers to get a third array of numbers. Used in image processing to stretch or resize images.

**Corrector plate** – The lens on the front of an SCT type telescope that corrects for the spherical aberration created by the spherical mirrors used in that design.

**Counterweight** – A weight, usually on an equatorial mount, used to balance the weight of the telescope and associated hardware.

**Crayford focuser** – A telescope focuser that uses smooth bearings and rollers as opposed to gears used in rack and pinion style. They usually come in dual speed (coarse and fine adjustments) and can have adjustable tension.

**CRW/CR2** - Canon's RAW image format.

**Dark frame** - An image taken at the same ISO, shutter speed and temperature as the light frames but with the lens cap/scope cap on, or the shutter closed. This is used to detect the thermal signature of the camera's sensor at these setting so they can be subtracted from your final image.

**Dead pixel** - Opposite of a hot pixel, a pixel that is stuck in the off position and registers as black regardless of the amount of light applied.

**Declination (DEC)** – Celestial coordinate measured from the celestial equator north and south of that line, from +90 degrees to the north to -90 degrees to the south, zero being the celestial equator.

**Deconvolution** - A method of image enhancement that corrects for the bad effects of convolution. This can substantially increase fine details in an image.

**Dew heater** - Usually a strip that heats up and is wrapped around a telescope near the optics. This warms the optics and prevents dew from forming.

**Dew shield** - A device attached to the end of a telescope and is like a hollow extension of the telescope tube. This delays the objective from collecting dew, and reduces the intake of extraneous light sources.

**Diagonal** - A device that has a mirror inside and reflects the image at a 45 degree or 90 degree angle for easier viewing. One side goes into the focuser, the other end holds an eyepiece.

**Diffraction** - As light passes through a telescope it passes through openings. As light gets near the edges of these openings it is diffracted. This causes stars to appear larger than they actually should.

**Diffraction limited** – Term used primarily by telescope manufacturers that says that the telescope should perform so that any defect seen will be with the physical characteristics of light and not optical problems with the telescope.

**Dispersion** – Cause of chromatic aberrations. Prism effect, when light is spread out into its spectrum from white light.

**Dobsonian** - a type of telescope mount, but usually used as a reference to the entire telescope assembly. These are usually larger Newtonians mounted onto a base that sits on the ground and moves as an alt/az. Like regular Newtonians these are not well suited to astrophotography due to not having enough backfocus.

**Doublet** – A refractor telescope with two objective lenses.

**Dovetail** - A metal rail that attaches to the bottom of the telescope, usually by rings that clamp into the telescope tubes or bolts into the bottom of the telescope, which can then be quickly and easily attached to the mount's clamp. Popular dovetail types include Vixen and Losmandy.

**DSLR** - Digital Single Lens Reflex camera. A type of camera where the user actually looks at the same image that will be recorded on the sensor by means of a mirror and prism that reflects the light from the lens through an eyepiece. When the shutter is opened to take the picture the mirror swings out of the way, the eyepiece goes black as it is no longer receiving the reflected image, and the sensor is exposed.

**DSS** – Short for Deep Sky Stacker, very popular free program generally used by beginning astrophotographers for stacking images.

**Dynamic range** - The range from brightest to darkest that a camera can record.

**ED** – Extra low Dispersion, optical glass corrected for chromatic aberration.

**EQ/Equatorial Mount** - A type of mount specifically designed to track the stars as they travel around the earth compensating perfectly for their arc in the sky.

**Ephemeris** – Detailed positional information about planets, their moons, comets and asteroids.

**Eyepiece** - An optical device that focuses the light exiting a telescope tube in such a way that you can view it with your eye. These typically contain many lens elements in a round cylinder that is inserted into the focuser. The eyepiece can be made to magnify or reduce the image size.

**Eyepiece projection** - A method of taking a photograph through the eyepiece of a telescope without a lens on your camera. This uses a specific adapter. This can come in handy on telescopes that cannot reach focus using a prime focus adapter.

**F-Stop** - When using a camera with its lens installed, the aperture is adjustable and is commonly referred to as the F-Stop.

**Field flattener** - An optical device used primarily on refractors to make sure that the image arrives at the camera sensor perfectly flat. This prevents elliptical images of stars in the corners of the images while the stars in the center may be perfectly round.

**Field of view** - Commonly represented as FOV. The area of the sky that you can see at one time. Longer focal lengths (more magnification) generally show smaller areas of the sky and hence a smaller field of view. Eyepieces with smaller numbers cause the same effect.

**Field rotation** – The effect of the image being blurred from the rotation of the sky. This can happen when you use an Alt/Az mount to take long exposures since the Alt/Az mount does not rotate the camera like an EQ mount does.

**Filter** – A filter is a piece of glass (or Mylar in some solar filters) that alters the light coming through the telescope before the eyepiece or camera. A filter is used for removing light pollution, enhancing certain colors, shooting color images with a monochrome camera and many other tasks.

**Finder** - A small telescope or other pointing device that helps you quickly orient your telescope towards a particular target. Similar to a gun sight.

**Firmware** - The software a device uses to tell it what to do. For example, your GoTo telescope software in the hand controller is called its firmware and can be updated on many devices.

**FITS format** - A file format designated by .FIT (such as .TIF, .GIF or .JPG) specifically designed for scientific purposes. Like RAW or TIF files this stores raw data that does not degrade from repeated editing as do formats such as .GIF or .JPG.

**Flats/Flat frame** - An image taken with even illumination over the front of the telescope and exposed to present a neutral gray image. This must be taken with the exact same setup as your light frames (same focus setting, same filters, etc) and is used to remove vignetting.

**Focal length** - The length of a line following where the light travels through a telescope, this is important for calculating parameters such as the FOV and magnification.

**Focal plane** – An inferred plane at the point where the image from the telescope comes to focus. A camera's sensor is mounted so that it is at the focal plane.

**Focal ratio (FR)** – The focal length divided by the aperture of the primary objective of the telescope.

**Focal reducer** - An optical device which reduces the effective focal length and increases the field of view of a telescope, seemingly reducing the magnification. This is usually mounted into the focuser before any eyepieces or cameras.

**Focuser** - A piece of equipment mounted on the telescope where the light exits. Eyepieces, diagonals, barlows and cameras are mounted into the focuser. Its job is to move the eyepiece/camera/etc back and forth until the light comes into focus at a specific point (your eye or the camera sensor).

**FOV** – See field of view.

**Frames per second (FPS)** – The number of image frames captured per second by the device, used in video capture devices.

**Full well capacity** - A measurement of the total amount of light a photosite can store before saturation occurs.

**FWHM** – Full Width Half Maximum. The measurement of the angular apparent size of a star, usually used to get the size as small as possible in an image which represents the best possible focus.

**Gain** - This is a multiplication of the incoming signal. For example, if one photon enters a camera and hits the sensor, setting the gain to 2x will cause the digital signal sent from the camera sensor to say that two photons hit the sensor. Increasing the ISO of a digital camera is increasing the gain.

**German equatorial mount (GEM)** – Another name for the equatorial mount.

**GoTo** – A telescope that when properly aligned can point to a celestial object automatically when selected from a catalog or menu.

**GPS** – Global positioning system, a device or feature used to determine your exact location on the planet.

**Grayscale** - An image recorded in black, white, and variations of gray with no color information.

**Guiding** - The act of following a star or other object using either manual corrections (as was the case back before GoTo and tracking mounts) or automatically using guiding equipment such as an autoguider.

**Hand controller (HC)** – The handheld device used to control your telescope's mount.

**HDR** - High Dynamic Range. You can use different exposures on different images and sandwich them together to show an image that has too much dynamic range to be captured in one single exposure. M42 is a prime example of a target that needs HDR processing: if you expose correctly for the faint dust lanes on the outer areas, the central core is blown out or clipped; if you expose for the central core, the outer dust lanes are clipped into blackness and can not be seen.

**Highlights** - Areas of maximum brightness in an image.

**Histogram** - A graph that shows how an image is exposed. In a normal grayscale histogram the left side is absolute black, the right side is absolute white and there is usually a hump in the graph display somewhere near the center showing the exposure of that image. Color works the same way but shows the intensity of the red, blue and green color channels.

**Hot pixel** - Opposite of a dead pixel. A pixel that shows exposure information even when shot in complete darkness.

**Illuminated reticle eyepiece** – An eyepiece with an illuminated crosshair or other centering marker used for precise centering of targets in the field of view.

**ISO** - International Standards Organization, used to measure the "speed" of film, or the sensitivity of a sensor in a digital camera. As ISO increases, less light is required to "expose" for a given image. This also reduces the signal to noise ratio, increases noise, and reduces the bit depth possible in the image.

**JPG/JPEG** - Joint Photographic Experts Group. A file format denoted by .JPG (such as .TIF or .GIF) that is very common in digital cameras. Using this format should be avoided because it uses a lossy compression format to reduce file size. This results in huge losses of information and makes it virtually impossible to process well for astronomical uses.

**Light frame** - A standard picture. Every regular picture you have taken with a regular camera of birthdays, friends and family are all what we call light frames. These are the frames you work with that contain your image data.

**Light pollution** – Stray light from street lights, signs, windows etc that shine or are reflected up into the air. This is scattered by contaminates and humidity in the air and create a glow effect around cities making it difficult to see outside the atmosphere.

**Light year** – The distance light travels in a year through a vacuum, approximately 5.87 trillion miles.

**Limiting magnitude** – The measurement of the dimmest star you can see at zenith which takes into consideration all parameters such as light pollution, weather conditions and optical devices used (if any).

**Lossless compression** - Certain file formats such as PSD and TIF employ compression methods that preserve 100% of the data while decreasing the file size.

**Lossy compression** - Formats such as .GIF and .JPG use lossy compression which throws away data that it does not think is needed to display the image.

**LRGB** - When shooting a monochrome camera and creating a color image you need to shoot at least one image with a red filter, one image with a green filter and one image with a blue filter. These are combined together into one color image. The L in LRGB stands for luminance and is used to increase detail in an image. The Luminance frame is the detail frame and can be shot in very high resolution. The color can be shot at lower resolutions and combined with the luminance to create a high resolution color image. You can use this idea to increase your ability to stretch images as well.

**Luminance** – The recording of brightness or intensity of light. Typically this is the high resolution/detailed portion of an image.

**Magnitude** - A measurement of the brightness of an object. An increase in one magnitude is approximately 2.5 times as bright. The lower the number on the scale, the higher the magnitude.

**Maksutov Cassegrain telescope** – See MCT below.

**Maksutov Newtonian** – Similar to a Maksutov Cassegrain except they are designed as a Newtonian configuration with the focuser near the front of the scope.

**MCT** - Maksutov Cassegrain Telescope, a type of telescope that has a sealed front end which is actually a corrector lens called a meniscus, two mirrors and has its eyepiece in the rear.

**Megapixel** - Roughly one million pixels.

**Meridian** - An imaginary line dividing the west and east halves of the sky running from the north celestial pole directly overhead to the south celestial pole.

**Meridian flip** - Meridian Flip is the act of re-orienting the scope on an EQ mount so it can continue to track past the meridian. This "flips" the scope around to pointing the other direction at roughly the same spot on the meridian. Going past the meridian without flipping can cause the scope to run into the mount, cables to come loose, and many other really bad things.

**Micron** – One millionth of a meter or 0.001mm.

**Mirror cell** – The frame that holds the primary mirror assembly.

**Mirror lock(DSLR)** – Some cameras have the ability to lock the mirror in the up position to minimize camera vibration when the shutter is tripped. This can be very useful shooting brighter objects like the moon but is ignored in long exposure work as the amount of time the camera is vibrating due to the mirror slamming open is miniscule compared to the overall exposure time.

**Mirror lock(SCT)** – Some SCT type telescopes have the ability to lock the mirror once the image is in focus to prevent the mirror from "flopping" or moving as the orientation of the telescope changes.

**Monochrome** – Technically means one color, meaning either black or white. "Monochrome" cameras are actually grayscale in that they produce black, white and many different shades of gray.

**Mosaic** - The act of shooting multiple images in a grid pattern and stitching them together to allow you to shoot a larger field of view than you could normally.

**Mount** - The mount is the geared (and sometimes motorized) device that is typically attached to the top of a tripod and then has the telescope attached to it. It is the mount that allows you to point the telescope at different objects without moving the tripod, and (when motorized) tracks objects across the sky.

**Narrowband** - Using special filters you can capture the emissions from certain gasses such as hydrogen alpha, sulpher and oxygen. These can be used much like LRGB imaging to create faux color images of high resolution. This method can also overcome all but the worst light pollution situations and can even allow you to shoot on nights with a full moon to some degree.

**Near Earth Object (NEO)** – An object such as a comet or asteroid which will pass in close proximity to earth.

**Newtonian** - A type of reflector telescope that has two mirrors in a hollow tube. The front of the telescope is open to the elements and the back is sealed. The eyepiece is near the front of the scope. These are usually not suitable for astrophotography unless they are designed as an "astrograph" as they will not bring a camera to focus without modifications or the use of a Barlow.

**North celestial pole (NCP)** – The point in space very close to Polaris where a line drawn from the exact southern to northern poles would extend into space with the earth revolving around that line.

**Nyquist theory** - States that when converting frequencies, the sampling rate should be 2x the highest frequency to get an accurate conversion and preserve all the data.

**Objective lens** – Also called the primary objective, the large front lens of a refractor telescope.

**Off axis guider (OAG)** – A method of mounting a guide camera so that it shares most of the same optical path as the imager, picking off a small amount of light usually from a mirror mounted in the light path.

**One shot color (OSC)** - Any camera that creates a color image from a single exposure.

**Opposition** – Opposition is when a planet is closest to the earth and is directly on the other side of earth from the sun.

**Optical train** - Anything that is directly in the path of light from the stars to your eye or camera sensor is considered "in the optical train". Could be called the optical path as well.

**Optical tube assembly (OTA)** – Also referred to as the OTA, this is the main tube of the telescope not including any mount, pedestal, pier or tripod.

**Parfocal** – Applies to both eyepieces and filters and means that if you exchange one filter (or eyepiece) for another, you will remain in nearly perfect focus. Not all filter sets or eyepiece sets are parfocal.

**Periodic error (PE)** - Errors in the manufacturing process of the gears and drive assembly in an EQ telescope mount results in repeating errors in the tracking of the mount. These can be removed with software that contains PEC code.

**PEC** - Periodic Error Correction. Software that corrects for periodic error.

**Photometry** – The measurement of apparent magnitude of objects such as comets, asteroids and stars.

**Photon** - For the purposes of discussion in this book, a photon is a single particle of light.

**Photosite** - The technical name for the tiny part of the sensor in a digital camera sensor that when exposed to light records a signal. Typically called a pixel.

**Piggyback** - Mounting a camera with a lens on a telescope in such a way as it is not shooting through the telescope but is instead just using it as a tracking mount.

**Pixel** - A single dot in an image.

**Pixel size** - The physical size of a photosite on the sensor of a camera, measured in microns.

**Plate solve** – Refers to Plate Solution, or finding the absolute position and motion of an object. Some applications such as TheSkyX Professional offer a plate solve feature where it can look at your image and tell you exactly what is in the frame.

**Point light source** - Stars are considered point light sources because regardless of their magnification they are so far away they will always appear as a single point of light.

**Polar alignment** – Aligning the "polar axis" of an equatorial mount to either the northern or southern celestial pole so that the mount can track celestial objects precisely.

**Polar scope** - A small telescope usually built into the mount which allows for precise pointing of the mount's right ascension axis to the north or south celestial pole.

**Prime focus** - Attaching a camera without a lens in such a way that the image from the telescope is directly projected onto the sensor of the camera.

**Quantum efficiency (QE)** - A measurement of the percentage of photons which hit a photosite versus how many are detected.

**Rack and pinion focuser** – A less expensive and typically less accurate style of focuser.

**RAW** - A RAW file is a file that contains the relatively unaltered, unmodified data directly from the camera's sensor.

**Rayleigh scattering** – The scattering of different wavelengths of light by the molecules in the atmosphere. This scattering is the reason the sky appears blue.

**Resolving power** – 4.56/(inches of aperture of the telescope)=resolving power of the telescope in arc-sec. Note that this does not take into consideration obstructions such as secondary mirrors.

**Reticle** - Crosshairs or other markings that allow you to precisely center a target in your field of view. Sometimes included inside eyepieces and finder scopes.

**Red dot finder** - A type of finder that uses an illuminated red dot as a reticle.

**Refractor** - A type of telescope that has an objective lens on the front end and an eyepiece or camera at the other. Light passes straight through without being reflected unless a diagonal is used.

**RGB** - Red, Green, Blue. One shot color cameras shoot everything as a combination of these three primary colors. When shooting monochrome images and wanting to end up with a color image, you shoot at least one frame with a red filter, one with a green, and one with a blue and then combine them to create a full color image.

**Right ascension (RA)** – Celestial coordinate measured from west to east in hours, minutes and seconds. As the earth turns each hour, 15 degrees of arc pass.

**Saturation** – The point at which you cannot record any more data. This may refer to the full well capacity of a CCD camera or the maximum value a pixel can store.

**Schmidt Cassegrain Telescope (SCT)** - a type of reflector that has a sealed front, two mirrors and has its eyepiece in the rear of the scope.

**Seeing** - A measurement of the conditions of the atmosphere as it relates to being able to view or image an astronomical object. An easy method to determine the seeing conditions is to look for stars twinkling; the more they twinkle, the worse the seeing.

**Sidereal rate** – 23 hours, 56 minutes and 4 seconds is one sidereal day which is why the stars are never at the exact same place at the exact same time every night and seem to "advance" across the night sky every night all year long. This is the rate at which your telescope must track to remain aligned with your target.

**Signal to noise ratio (SNR)** - The ratio of signal (what you are trying to capture in the image) to noise (electrical signals inherent to the camera generating the image). The higher the SNR, the easier it is to stretch an image and bring out the detail of your target.

**Slew** – The process of your telescope moving to and from targets.

**South celestial pole (SCP)** – The point in space very close to Sigma Octantis where a line drawn from the exact northern to southern poles would extend into space with the earth revolving around that line.

**Spider vanes** - Small strips of metal or plastic in the front of a Newtonian telescope which supports the secondary mirror in the optical path.

**Stacking** - Taking several images and combining them in such a way as to increase the signal that you want to keep while reducing the noise levels that you do not.

**Strehl ratio** – Gives a ratio as compared to a theoretically perfect optical system. For example, a Strehl ratio of .90 is 90% as good as a theoretically perfect optical system.

**Stretching** - Taking an image and manipulating the data so that details that were too dark to see are now light enough to be visible through compression of the grayscale or color scale.

**T-Ring** – An adapter that mates with a removable lens camera on one side and has threads on the other side to attach to the telescope or other device.

**Thermo Electric Cooler (TEC)** – Electric cooling device used with some CCD and DSLR cameras.

**TIF** - A file type (like .GIF and .JPG) to store image files. TIFs are excellent because they are lossless formats. They are however far larger than JPG or GIFs.

**Tracking** - The ability to follow an object as it appears to travel across the sky.

**TSX** - Abbreviation for TheSkyX, a planetarium, telescope control and planning application for amateur and professional use from Software Bisque Inc.

**United States Naval Observatory (USNO)** – The standard for timekeeping in the United States.

**Vignetting** - The effect of the edges of an image being darker than the center due to obstructions or optical imperfections.

**Well depth** - A measurement of the total amount of light a photosite can store before saturation occurs.

**White point** - A part of an image that represents pure white.

**Zenith** – The point directly overhead.

## 5.4: Other books by the author:

Want to take a few snapshots of the beautiful objects you are viewing without spending a small fortune? Already have a camera but you can't seem to get a good image and want to know why?

This book will answer those and many other questions while giving you a quick and reasonably easy introduction to budget astrophotography. In addition, save more money by seeing how to make a lot of items you may find useful.

http://www.allans-stuff.com/bap/

If you decide that you want more than quick snapshots, you want big beautiful prints to hang on your wall, this is the book for you.

From required and optional equipment, through the capture process and into the software processing needed to create outstanding images, this book will walk you through it all.

http://www.allans-stuff.com/leap/

# 50 Amazing Things to See With Your New Telescope

You decide that you want to take images of celestial targets, but need a little help with the targets? This book discusses all 110 Messier targets and includes descriptions, realistic images of each target, star charts and shoot notes to help you image all 110 of the objects yourself.

http://www.allans-stuff.com/mar/

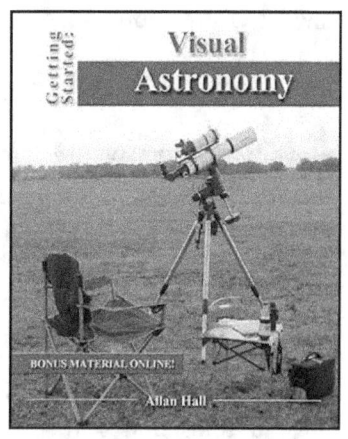

If you have ever wanted to view the wondrous objects of our solar system and beyond, here is the how-to manual to get you well on your way. From purchasing your first telescope, through setting it up and finding objects, to viewing your first galaxy, this book contains everything you need. Learn how to read star maps and navigate the celestial sphere and much more with plenty of pictures, diagrams and charts to make it easy. Written specifically for the novice and assuming the reader has no knowledge of astronomy makes sure that all topics are explained thoroughly from the ground up. Use this book to embark on a fantastic new hobby and learn about the universe at the same time!

http://www.allans-stuff.com/va/

Many midrange and high end telescopes come on equatorial mounts. These mounts are fantastic for tracking celestial objects. Someone who wanted to take pictures of objects in the night sky might even say they are required for all but the most basic astrophotography. The problem is that they can also be unintuitive and require some knowledge to use.

If you have ever struggled to figure out how to use an equatorial telescope mount, this is the book you always wished you had.

http://www.allans-stuff.com/eq/

Have you ever wanted to take a picture of a solar or lunar eclipse but didn't know where to start? This is the book for you!

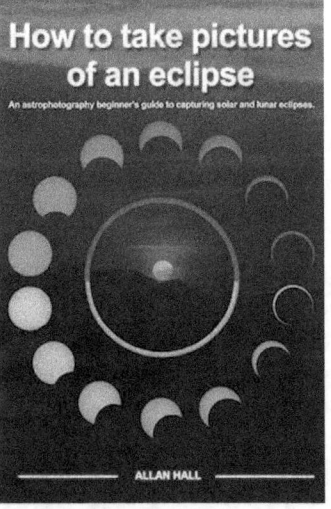

Whether you want to snap a few pictures with your smartphone or have a telescope with solar filters, this book will help you prepare, find the perfect location, and get those incredibly rare and beautiful images.

http://www.allans-stuff.com/eclipse/

The Dobsonian telescope is one of the most popular styles of telescopes for beginner to intermediate amateur astronomers out there, with good reason. These telescopes provide excellent views for a modest investment, and are also very easy to setup and use.

This book will make sure that before you make your investment you know which telescope meets both your needs, and budget. In addition, you will feel comfortable not only purchasing one, but using one as well.

http://www.allans-stuff.com/dob

Once you know you want to pursue astrophotography, how do you know which of the tens of thousands of possible objects your beginner equipment can take successful images of? How about which ones are the right size for your equipment? What are the raw images supposed to look like when they come out of the camera?

Written specifically for the beginning astrophotographer with beginner equipment such as a small refractor, small reflector, or similar telescope on an EQ mount using a DSLR camera, this book will help you start capturing stunning images quickly and easily.

http://www.allans-stuff.com/50best

How to Find Objects in the Night Sky- One of the worst feelings for beginning amateur astronomers is when you either can't find an object you are looking for, or have no idea if what you are looking at is the object you set out to find.

With step by step instructions coupled with easy to understand graphics, you will learn how to find any object in the night sky with certainty.

https://allans-stuff.com/find/

So you've decided to write a book and get into non-fiction publishing. Now you find yourself faced with the seemingly infinitely harder second step – actually bringing the idea to market. In today's brave new world of self-publishing and open creative markets, it is both an inviting and potentially intimidating arena for authors hoping to turn their non-fiction books into a meaningful source of income. This is a daunting task because it  involves a blend of several disciplines that aren't necessarily part of an author's quiver of arrows. Most crucial among these are marketing and digital publishing, each of which requires fluency in fields that authors may or may not have experience in.

http://www.allans-stuff.com/ck/

# 50 Amazing Things to See With Your New Telescope

WordPress is the perfect tool to help you build the website you've always wanted. But the 'help' aspect which is built into it isn't always the right thing for someone who just getting started.

What you need, and what this book will provide, is a book that shows you how to get off the ground and then build on that knowledge to give you a secure and usable website.

http://www.allans-stuff.com/wp/

Information Technology is an area which is constantly on the move, sometimes at a speed which is dizzying and difficult to keep pace with. In particular **data recovery** can be one of the more complex problems you might encounter. The sheer amount of information is often overwhelming and confusing.

*Data Recovery* for Normal People is a new book which aims to make this process a lot simpler. Designed for both beginners who have little knowledge of technical issues and for those who may own their own computing business and want to learn more.

http://www.allans-stuff.com/dr/

## 5.5: NOTES:

**NOTES:**

## NOTES: